Frances V. Hickson
Roman Prayer Language
Livy and the Aneid of Vergil

Beiträge zur Altertumskunde

Herausgegeben von
Ernst Heitsch, Ludwig Koenen,
Reinhold Merkelbach, Clemens Zintzen

Band 30

Springer Fachmedien Wiesbaden GmbH

Roman Prayer Language
Livy and the Aneid of Vergil

Von
Frances V. Hickson

Springer Fachmedien Wiesbaden GmbH 1993

ISBN 978-3-663-12347-7 ISBN 978-3-663-12346-0 (eBook)
DOI 10.1007/978-3-663-12346-0

Die Deutsche Bibliothek – CIP-Einheitsaufnahme

Hickson, Frances V.:
Roman prayer language: Livy and the Aneid of Vergil /
von Frances V. Hickson. – Stuttgart: Teubner, 1993
(Beiträge zur Altertumskunde; Bd. 30)
Zugl.: Chapel Hill, Univ. of North Carolina, Diss., 1986 u. d. T.:
Hickson, Frances V.: Voces precationum
ISBN 978-3-663-12347-7
NE: GT

Ursprünglich erschienen bei B. G. Teubner Stuttgart in 1993
Softcover reprint of the hardcover 1st edition 1993

magistris meis

Table of Contents

Abbreviations

The names of ancient authors and their works are normally abbreviated according to the conventions of standard English reference works; these abbreviations may be found in the *Index Locorum*. Abbreviations for periodicals follow those of *l' Annee Philologique*. Secondary literature is generally cited by the author's last name and the date of publication if necessary for clarity. Full references appear in the Bibliography. Other abbreviations are as follows:

Lewis and Short=Lewis, Charlton and Short, Charles, ed. *A Latin Dictionary*. Oxford 1975.

Mommsen *StrR.*=Mommsen, Theodor. *Römisches Strafrecht*. Leipzig 1899; reprint Graz, 1955.

Mommsen *StR.*=Mommsen, Theodor. *Römisches Staatsrecht. Handbuch der römischen Alterthümer*. Vols. 1-3. Leipzig[3] 1887-1888.

OLD=Glare, P. G. W. ed. *Oxford Latin Dictionary*. Oxford 1982.

RE=Pauly-Wissowa's Real-encyclopädie der classischen Altertumswissenschaft. Stuttgart 1893- .

TLL=Thesaurus Linguae Latinae. Leipzig 1900-.

VIR= *Vocabularium Iurisprudentiae Romanae*. Berlin 1903-1987.

Preface

This study represents an extensive revision of my dissertation, *Voces Precationum: The Vocabulary of Prayer in Livy and the Aeneid of Vergil*, accepted by the University of North Carolina at Chapel Hill in May 1986. The study has been expanded to include chapters on vows and oaths.

I dedicate this work to the many teachers who have inspired, guided, and encouraged my study of classical languages and cultures over the years. Their number is too great to name all individually, but I wish to single out three individuals. I gladly acknowledge my greatest debt of gratitude to Jerzy Linderski, who has been my professor, dissertation advisor, mentor and friend. His support and encouragement of my work date back to the very first semester of my graduate studies and have never failed. I wish to address a special word of thanks to Laura V. Sumner and Lucille C. Jones, my professors of Classics at Mary Washington College in Fredericksburg, Virginia.

I also wish to express my appreciation to a number of individuals whose advice and assistance have been essential to the realization of this project. L. Koenen, editor of this series, offered helpful suggestions on content and patiently advised me on the preparation of camera-ready copy. Others who took time out from their own work to read my manuscript and offer comments were my colleagues A. Bennett, B. Jordan, and R. Renehan. My graduate assistants W. Morison and D. Dumont spent many hours proof-reading and formatting, not to mention their numerous trips to the library. S. Gardner contributed her journalistic experience in editing. F. Nennig and K. Carver, members of the staff of the UCSB Department of Classics, contributed valuable assistance.

Several institutions provided support as well. The National Endowment for the Humanities and the David and Lucille Packard Foundation provided funds for the APA's Fellowship in Latin Lexicography. During my tenure at the Thesaurus Linguae Latinae (1986-1987), the General Director, P. Flury and others provided rigorous training in lexicographical method. All those who have had the priviledge of access to the TLL archives and library will appreciate what this means to a scholar working in our field. The University of California at Santa Barbara generously granted a Faculty Career Development Award for two quarters of sabbatical leave in 1988-1989. During that sabbatical, the authorities of the University of North Carolina at Chapel Hill graciously appointed me a Visiting Scholar with access to all libraries and services. My former professors in the Classics Department there welcomed me with collegial hospitality and assistance. I single out the fact that T. R. S. Broughton and J. Bolter made the use of their personal offices available to me. The staff, especially D. Stolle, guided me through various bureaucratic labyrinths.

My gratitude to friends for support and encouragement cannot adequately be expressed. I should like to mention in particular the Shugart family who opened both their home and their hearts to me during my sabbatical and subsequent visits to Chapel Hill. Others who shared in this project include C. de Catanzaro, G. Commins, J. Cromartie, S. Gardner, R. Hahn, D. Keller, C. Konrad, A. Mallard, L. Moes, T. Moore, T. Papillon, and J. Rudestam.

Santa Barbara, California

August 1992

Ex homine remediorum primum maximae quaestionis et semper incertae est polleantne aliquid verba et incantamenta carminum.
Pliny, *Historia Naturalis* 28.10

A Prologue to Roman Petitionary Prayer

For the ancient Romans, the wording of prayers was a matter of crucial importance. In response to his own question, "do ritual words and incantations possess any power?" Pliny adduces numerous examples of the scrupulous use of prayers and incantations at every period of Roman history and in all areas of life to show that the Romans, in his opinion, both publicly and privately, maintained a genuine belief in the power of ritual words (28.10-11). Roman cult, especially that of the state, set high value on the careful repetition of precise formulae, which long experience had shown to be an effective means of communicating with the divine beings whose good-will was essential for the safety and prosperity of the state and its citizens. In their private lives as well, the Romans used numerous religious formulae to express their thoughts and feelings. There were formulae for capturing the attention of the gods, declaring the purpose of prayers, setting forth requests, and identifying beneficiaries. In the state cult, much effort was invested in practices that assured the preservation and correct repetition of traditional formulae. At the same time, the necessity for precise identification of divinities and requests required adaptability to varying situations. The study of prayer language shows both synchronic and diachronic variations which evidence this quality.

It was not only in the composing and speaking of prayers that the Romans manifested their faith in the power of language. The modern reader will more readily understand such a belief concerning the composition of literature. Language has the power to capture the attention of a human audience, as well as a divine one, and to shape its response to a work of literature. The fundamental process of writing is the choice of words and their arrangement. By setting familiar language in familiar contexts, an author summons up in the audience clusters of images attached to previous encounters with that language in writing or speech.

The use of words in new or unusual arrangements and contexts evokes fresh responses, mingling images associated with different settings.

The power of such authorial choices is most obvious when the text under examination is one in which the surrounding culture dictates the use of traditional language, as in ancient Roman prayers. When literary authors employ traditional religious formulae, they bring a wealth of communal experiences to their texts. Alternatively, they can borrow language from literature, rather than from life, thus creating an interplay between cultural and literary experiences. Or if authors choose to do so, they can select words and arrange them so as to compose entirely new phrases and to create new images. The study of such literary choices in specific texts offers insights into the cultural and literary environment of those texts and into possible audience response.

Two Roman authors, Livy and Vergil, provide an especially good opportunity to examine the creative choice between formulaic language and literary interpretations in the context of prayers. Both Livy's history and Vergil's *Aeneid* contain numerous prayers of varying type and content. Both works share the cultural context of the Augustan period. Furthermore, both authors are reputed authorities on Roman religion. Modern scholars often cite the prayers of Livy's history as authentic or at least as modernized versions of archaic prayers.[1] The ancient Romans themselves proclaimed Vergil's expertise; in Macrobius' *Saturnalia*, Praetextatus goes so far as to call Vergil *pontifex maximus* (1.24.16).[2]

The present study focuses on the language of prayers in these two works.[3] The approach is lexical and its objective is two-fold: to show the

---

1 For example, Fowler writes, "in two or three cases Livy has copied the formula from the *tabulae* of the pontifices" (204). On the *tabulae*, see Frier, esp. 83-105.

2 On ancient opinions of Vergil, see Lehr 11, 15-16. Adde Macrob. Sat. 3.2.1, 7, 10.

3 The essential foundation for the study of Roman prayer is Appel's *De Romanorum Precationibus*. Appel sometimes demonstrates a too ready acceptance of prose texts, particularly Livy, as sources for authentic Roman prayers. From such a perspective, it is not surprising that he neglects Livy's phraseology when not well attested as technical, e.g. *arcere* (126). For more specialized studies, I refer the reader to the works of Block, Gagnér, Highet, Jeanneret, Lehr, Ogilvie (1969), Secknus, Shatzman, and Swoboda-Danielewicz. In addition to these author-specific studies, there are detailed investigations of certain religious words and word families: Cipriano on *fas*; Erkell, Fugier, and Zieske on *felix*; Schilling (1954), Versnel (1981), and others on verbs of petition. There is also the thorough article by Alvar (1985) on the formula *sive deus, sive dea*.

insights which that language gives into one important aspect of Roman religion and into the writings of these two great Roman authors. This prologue provides a general introduction to prayers in Roman religion. The first chapter considers the relationship between prayers in Livy's history and Vergil's *Aeneid* and those found in their predecessors. Discussions of the content and conceptual bases of petitionary prayers appear in chapters two through five. For the reader who is interested in the particular language of prayers in Livy and Vergil, whether from the perspective of a classical philologist or an historian of religions, these chapters continue with a lexical analysis of specific words and phrases characteristic of Roman cultic prayer. Also included is language of a purely literary origin which the authors, especially Vergil, used in lieu of technical phraseology. Chapter six considers a small number of prayers that occur in similar contexts in Livy and Vergil and compares their language. The study concludes with a general summary of observations made in previous chapters about Livy and Vergil's usage of individual words and formulae. While my main interest is Roman religion and my chief approach is lexical, I hope, nevertheless, that some of my materials and results may be of interest to literary historians.

Because of the size of the topic, I have limited this study in several ways. I have restricted analysis of the historical development of language almost solely to usage in prayers and only occasionally mentioned related religious uses, such as augural; there has been no attempt to provide a broad analysis of each item in non-prayer contexts. The study does not include nouns or adjectives occurring only in epithets, such as *pater*, which, despite implicit religious concepts, function primarily as respectful terms of address. I have also limited the types of prayer examined. Although the English word "prayer" has come to indicate several different types of speech addressed to a god or gods (adoration, confession, lamentation, petition, and thanksgiving), only petition appears with frequency among the surviving prayers of Roman cult. Because of this prominence and in order to maintain some degree of topical unity, I have restricted this study to petitionary prayers, including the sub-types of vow and oath. Other types such as thanksgiving and dedication must await a separate study.

Petitionary Prayer in Roman Religion

Prayer, that form of human speech which explicitly or implicitly addresses immortals, was the most ubiquitous phenomenon of Roman religion, more so even than sacrifice since in addition to accompanying all sacrifices and other rituals, it also existed as an independent act. As an accompaniment of rituals, it described and explained the acts by which the Romans sought to ensure the *pax deum*, "peace with the gods." Prayer indicated for whom the ritual was intended and described its contents, for example, a libation or the sacrifice of an animal. It also indicated the purpose of the ritual such as expiation, propitiation, or aversion of evil. As an independent phenomenon, prayer addressed some of the same purposes as rituals but could do so in more varied circumstances. A vow could be announced in the midst of battle; an apotropaic exclamation, in the midst of conversation. Furthermore, not every situation required a ritual action; a prefatory formula, for example, sufficed before an assembly of the people.

The most common type of prayer among the Romans, as well as that most illustrative of the spirit of their religion, was the petitionary prayer, simply defined as a request made by a human being to an immortal. The Latin words *precor, preces,* and *precatio,* which English speakers regularly translate as "pray" and "prayer," are essentially words of petition. The basic Roman concept of religion was practical, concerning itself with obtaining earthly "peace" and prosperity. Among the various petitionary prayers attested in ancient Rome, two variations stand out as distinctive by virtue of their language: the vow, a promise conditional upon the fulfillment of a petition, and the oath, a petition for witnesses and the punishment of perjury. Therefore, these two sub-types each receive separate discussions in this study.

Occasions of Petitionary Prayer

Daily life, public and private, in ancient Rome presented many occasions for petitionary prayer.[4] The most obvious occasions were those marked as sacred by their overt religious associations. These included regular cultic observances with their varied rituals. They might be major public events such as the festival of the Parilia or minor domestic ones

4 See Appel 56-63 for detailed examples of prayers including vows and oaths.

such as the offering before a meal. Proximity to a sacred place, a temple or shrine, might also evoke a spontaneous private petition. Other sacred occasions were concerned with divination. On official occasions, as at the inauguration of Numa, the legendary second king of Rome, specific petitions requested the manifestation of divine will (Liv. 1.18.9=App. 5). Alternatively, if unsolicited signs appeared, such as the vision of Mercury that addresses Aeneas in a dream, observers responded with petitions for the fulfillment of favorable omens or the aversion of unfavorable ones (*Aen.* 4.265-278).

In addition to these clearly sacred occasions, some apparently profane experiences elicited petitions for the manifestation of divine power. Frequently the situation was an ambiguous one, a situation that could have either positive or negative outcomes. The petitioner asked that the divinity exert its power for the aversion of harm and for the procurement of success or prosperity. Such an occasion especially existed at times when some new activity or undertaking was about to commence, for example, civil meetings, wars, journeys, or marriages. At other times, the feared danger was immediate, as in the face of military rout, famine, or disease.

Speakers

The persons who spoke these prayers varied according to the setting, especially as distinguished by public or private occasions. Public religion was in the hands of representatives of the state, both religious and civil. In certain cultic observances, the priests of individual deities (*flamines*) offered appropriate prayers. But when the ceremony was of overtly public interest, such as a vow for the welfare of the state, a magistrate would speak words that were dictated to him by a *pontifex*. On some occasions, both officials and private citizens offered prayers together. In times of danger, present or anticipated, the senate could ask that all men, women, and children address prayers to the gods in what was probably the most dramatic of these cultic occasions, the collective prayer (*supplicatio*) before the altars of all the gods.[5] In domestic cult, the *paterfamilias* normally offered prayers for the welfare of the entire household, for example, those prayers which, along with an offering, preceded the chief

---

5 On the three types of *supplicatio* (expiatory, propitiatory, and gratulatory), see Halkin 9-13 with literature.

daily meal. When necessary, however, the *paterfamilias* could ask someone else to stand in for him, as an absentee landlord might ask a steward to do (e.g. Cato *Rust.* 141). Individuals could, of course, always offer their own private prayers in personal situations. Brief exclamatory prayers, especially oaths, colored the everyday speech of many Romans.

Orality of Prayers

Ancient prayers were spoken aloud or silently or murmured under the breath, depending primarily on the purpose of the speaker.[6] Naturally, prayers on behalf of a community, whether the state or a private cult, were generally recited aloud. Thus the vows announced by a general before departing on a military expedition (*profectio*) would be spoken loudly and clearly before a crowd of his troops and onlookers. So too, prayers offered on behalf of an estate with its human occupants would be spoken aloud so that the participants in the ceremony might hear them. Even solitary prayers of persons making individual requests were often spoken aloud.

There were other occasions when silent or quiet prayer was desirable, such as private requests relating to health or heart. In exceptional cases, public priests too had occasion to pray silently; Plutarch mentions such at the entombment of an unchaste Vestal (*Num.* 10.4-7). The Romans, however, generally associated murmured prayers with malevolent purposes. Horace dramatizes the point well:

> Vir bonus, omne forum quem spectat et omne tribunal,
> quandocumque deos vel porco vel bove placat,
> "Iane pater!" clare, clare cum dixit "Apollo!"
> labra movet metuens audiri: "pulchra Laverna,[7]
> da mihi fallere, da iusto sanctoque videri,
> noctem peccatis et fraudibus obice nubem"
> (*Epist.* 1.16.57-62).

If prayers of this sort were too embarrassing to be said aloud, how much more so magical spells for evil purposes.

---

6 On the orality of ancient prayer, see Versnel (1981) 25-28 with literature; Wagenvoort (1980) 197-209; Appel 209-212. On incantations, see Sudhaus 197-199.

7 Laverna was a goddess of theft.

The Power of Words

The hallmark of Roman cultic prayer was concern for the precise formulation and repetition of ritual language.[8] For the encyclopedist Pliny the Elder, this concern provided evidence of the trust that people placed in the efficacy of "ritual words and incantations." Describing the state cult, he writes:

> Ex homine remediorum primum maximae quaestionis et semper incertae est, polleantne aliquid verba et incantamenta carminum? Quod si verum est, homini acceptum fieri oportere conveniat, sed viritim sapientissimi cuiusque respuit fides, in universum vero omnibus horis credit vita nec sentit. Quippe victimas caedi sine precatione non videtur referre aut deos rite consuli. Praeterea alia sunt verba inpetritis, alia depulsoriis, alia commendationis, videmusque certis precationibus obsecrasse summos magistratus et, ne quod verborum praetereatur aut praeposterum dicatur, de scripto praeire aliquem rursusque alium custodem dari qui adtendat, alium vero praeponi qui favere linguis iubeat, tibicinem canere, ne quid aliud exaudiatur, . . . (*HN* 28.10-11).[9]

For recurring situations, the language of an entire prayer was fixed (*certa verba*). On other occasions, which varied in particulars, but maintained certain consistent elements, the language could be modified to fit the specific situation (*concepta verba*).[10] Thus, Livy presents the formula for demanding reparations from a foreign state with an indefinite phrase where the name of a particular party should be specified: "*audi, Iuppiter,*" *inquit*; "*audite, fines*"—*cuiuscumque gentis sunt nominat;*— "*audiat fas*" (1.32.6=App. 142). The proper wording of prayers used in the state cult was preserved in priestly books, from which the prayers were read on ritual occasions (cf. Cic. *Dom.* 140; Gell. *NA* 13.23.1). This was particularly important in the case of archaic prayers that the speakers

---

8 For the ancient references on this subject, see Wissowa 396-398. On the ritual gestures accompanying prayer, see Appel 184-214; Sittl 174ff.

9 On this passage, see Köves-Zulauf, esp. 24-63. On commendation as a type of prayer, see the review of this book by Linderski.

10 For this terminology, see Linderski (1986a) 2266-2267 with n.476.

themselves no longer understood, as Quintilian remarks of the archaic Salian hymn (*Inst.* 1.6.40).[11]

In spite of all precautions, mistakes could and did occur in the recitation of prayers. Livy, for example, tells of an occasion when the Latin Festival was invalidated because "in one sacrifice the magistrate from Lanuvium had omitted the words *populo Romano Quiritium*" (41.16.1). As a result, the ceremonies had to be renewed (*instauratio*) and the people of Lanuvium offered new sacrifices (cf. the Iguvine tablets 6B 48). Obviously, the Roman state took this matter very seriously; omission of the name of the Roman people would mean that they would not receive the benefits requested in the prayer. Cicero mentions other flaws in prayer that could invalidate religious proceedings in his speech *de Domo Sua*. Here he asks the pontiffs to declare invalid the consecration of his house performed by his enemy Clodius. He cites the inexperience of a young priest, who dictated the formulae of consecration *mente ac lingua titubante*. He also criticizes the faulty repetition of those formulae by the magistrate Clodius, who *praeposteris verbis, ominibus obscenis, identidem se ipse revocans, dubitans, timens, haesitans omnia aliter ac vos* (*sc. pontifices*) *in monumentis habetis et pronuntiarit et fecerit* (*Dom.* 139, 140).

The concern with language in cultic prayer went beyond precise recitation of set prayers and extended to the scrupulous composition of formulae. Here religious anxiety before the unpredictable power of the gods led to linguistic attempts to eliminate some of the potential mistakes that could cause petitions to go unanswered or to be answered otherwise than intended. The resulting language bears certain similarities to legal diction. In vows, there appear precautionary statements to ensure the validity of votive offerings despite unintentional mistakes in procedure, for example, *si atro die faxit insciens, probe factum esto* (22.10.6=App. 123). The same religious anxiety brought about the piling up of synonyms throughout prayers, for example, *precamur o[r]amus obsecramusque* (CIL 6.32329.13=Pighi 157.13) or *quod bonum faustum felixque sit* (CIL 6.30975=ILS 3090). In the invocation of the gods, this concern created a multiplication of divinities, epithets, descriptive predicates, and sometimes generic substitutes such as *vel quo alio*

---

11 E.g. *cozeulodorieso. omnia vero adpatula coemisse. | ian cusianes duonus ceruses dunus Ianusve | vet pom melios eum recum* (Morel, FPL *Carm. Sal. frg.* 3). The hymn of the Arval brotherhood also contains obscure phrasing (Henzen CCIV).

nomine te appellari volueris (Serv. *auct. ad Aen.* 4.577), all in order to be certain that the correct god was addressed.[12]

Despite the clearly strong prejudice toward preservation of set phrasing in prayers, there is also evidence of linguistic flexibility. The story of Scipio Aemilianus' alteration of the traditional prayer at the close of a census illustrates this quality (Val. Max. 4.1.10). When the scribe dictating the official prayer reached the petition for increased prosperity and growth of the state, Aemilianus refused to make such a request; the state was sufficiently prosperous and large. He immediately altered the prayer to request the continuance of the current state of affairs. The new version was duly recorded in the priestly books. It has been noted that, as it stands, the story is impossible: such an event would have invalidated the entire ceremony. Nevertheless, the underlying theme of changing religious formulae is certainly correct. Perhaps there was a procedure for alteration prior to a ceremony.[13] Such openness to change is also revealed by the existence of multiple formulae for certain occasions. In some cases, these variations occur concurrently; in others, the available evidence points to changes occurring over time.

Returning to Pliny and his question about the efficacy of prayer, we read *si quis (sc. precationem quandam) legat, profecto vim carminum fateatur, ea omnia adprobantibus DCCCXXX annorum eventibus* . . . (*HN* 28.12). This brings us back to an earlier point, the practicality of Roman religion. Their religious system was, in the opinion of Romans, based on empirical proofs. If their prayers were answered, they had obviously used the proper words and rituals; thus, the prayers were preserved to be used over and over again. If requests went unanswered, a mistake had somehow been made. The job then was to discover the mistake and correct it; when that was done, however long it might take, the gods would again listen favorably to the Romans' prayers.

Structure of Prayers

As a means of illustrating the typical structure of a simple petitionary prayer, an authentic prayer offered by Caesar Augustus at the *Ludi*

12 On the varied and numerous elements of invocations, see Norden (1913) 143-176 and Appel 75-114.

13 I follow the suggestion of North (3) after Bilz. For discussion and literature on this passage see also Harris 118-120, esp. 119 n.3.

Saeculares of 17 B.C. provides an interesting Augustan model for a general discussion.

> Moerae! uti vobis in illeis libri[s scriptum est, quarumque rerum ergo quodque melius siet p. R. Quiritibus, vobis IX] | agnis feminis et capris femi[nis propriis sacrum fiat: vos quaeso precorque, uti imperium maiestatemque p. R.] | Quiritium duelli domique au[xitis, utique semper Latinus obtemperassit, incolumitatem sempiter] | -nam victoriam valetudine[m p. R. Quiritibus duitis, faveatis p. R. Quiritibus legionibusque p. R.] | Quiritium, remque p. populi R. [Quiritium salvam servetis maioremque faxitis, uti sitis] volentes pr[opitiae p. R.] | Quiritibus, XV virum collegi[o, mihi, domo, familiae, et uti huius] sacrifici acceptrices sitis VIIII agnarum | feminarum et VIIII capraru[m feminarum propri]arum inmolandarum; harum rerum ergo macte hac agna femina | inmolanda estote, fitote v[olente]s propitiae p. R. Quiritibus, XV virum collegio, mihi, domo, familiae (CIL 6.32323.92-99=Pighi 114.92-99).

The prayer opens with an invocation (*Moerae!*), naming the deities whom the speaker addresses. Since the speaker knows the correct name of the invoked deity, the prayer begins simply with that name. Here, if necessary, would be added epithets or phrases to make the invocation specific. Next, the petitioner describes the sacrifice to be offered and its general objective of divine favor. Then, verbs of petition (*quaeso precorque*) express the intention to make a request. The specific favors requested follow (*auxitis, obtemperassit, duitis, faveatis, servetis, faxitis*). These include the common request for divine favor (*sitis volentes propitiae*). There follows a request that the sacrifice be accepted and again that the deities be *volentes propitiae.*

Such references to sacrifices are the typical Roman means of persuading the gods to listen favorably to a petitioner's request. As such, they are examples of a feature of prayers sometimes termed the argument, which presents the reason why the petition merits a favorable response.[14] In authentic Roman petitions there are a limited number of persuasive arguments used; persuasion generally consists of either mention

14 The term argument, used by Bremer, seems preferable to the traditional *pars epica* as broader (196 and n.15).

of an accompanying sacrifice or the vow of a future offering. Extant authentic Roman prayers differ from many prayers in Greek and Latin literature, especially in poetry, which make use of an argument based on past deeds of piety performed by the supplicant and previous favors granted by the divinity.[15] The *Aeneid* contains several examples of this sort of argument, including Nisus' words to Diana:

> si qua tuis umquam pro me pater Hyrtacus aris
> dona tulit, si qua ipse meis venatibus auxi
> suspendive tholo aut sacra ad fastigia fixi . . .
> (9.406-408).

Most of the prayers found in Livy and Vergil are not so complete as the Secular prayer to the Fates; they may consist, for example, only of one request. But taken as a whole, Livy's history and the *Aeneid* furnish examples of the language characteristic of each of these sections, which will provide the basic framework for a more detailed examination of concepts and language in later chapters.

Ancient Sources

The necessary foundation for any study of Roman prayer is a body of authentic prayers, as distinct from those that are the product of literary creativity. Although relatively small, the number of likely authentic prayers available is sufficient for a basic understanding of their language and structure. They furnish comparative material with which to examine the language of prayers found in Livy and Vergil.

Inscriptions provide examples of religious language from a broad time period. From the Republic there survive a number of inscriptions recording dedications to divinities identified only as *sive deus sive dea.*[16] Altars at Narbo, Salona, and Tibur preserve the wording of the religious law for the Republican altar of Diana on the Aventine (CIL 12.4333=ILS

15 For the traditional view that the reference to the past is a part of Roman prayer, see Ogilvie (1969) 36-37; Fraenkel (1957) 173; Appel 149-152. Notice that Ogilvie (1969), in a translation of this same prayer, incorrectly interprets *auxitis* as a reference to the past actions of the gods (31-32).

16 See Alvar (1985) 237-248, for a commentary on inscriptions containing this formula.

112; CIL 3.1933=ILS 4907; II 4.1 n. 73).[17] These inscriptions share certain words and formulae with official prayers.

Of particular interest in the study of prayers in Livy and Vergil are the inscribed records of prayers offered at the Augustan revival of the *Ludi Saeculares* in 17 B.C. (CIL 6.32323.92-146=Pighi 114-117.146).[18] Although these inscriptions are heavily damaged, records of the Severan games enable significant restoration of their Augustan models (Pighi 157ff.=CIL 6.32329). The precise value of the Augustan prayers as evidence for dating religious language is controversial. Despite their alleged source in earlier celebrations of these games, these prayers contain certain phrases which indicate a more recent date of composition or at the very least modernization. Nevertheless, some of their language is definitely confirmed by earlier sources.

Imperial inscriptions recording the rites of the *Fratres Arvales* preserve a linguistic tradition which dates back through the Augustan to the Republican period (*Acta Fratrum Arvalium*: CIL 6.2023-2119, 32338-32398; 37164-37165).[19] This apparently archaic priesthood, which died out during the Republic, was revitalized by Augustus. A large number of their records from the first centuries of the Empire have been found at the site of their cult. In an inscription from A.D. 218, the Brotherhood preserved what purports to be the *carmen Arvale*, an archaic hymn for the protection of crops (Henzen CCIV.32-37=CIL 2104.32-37). Annual prayers offered for the well-being of the emperor probably bear a close resemblance to those made for the well-being of the state during the Republic. These documents contain several religious formulae which appear in Republican prayers.

---

17 The leges ararum at Narbo and Salona include the statement *ceterae leges huic arae titulisque eaedem sunto, quae sunt arae Dianae in Aventino*. According to Dessau (ILS 112), the inscription at Narbo is Antonine, although the consular names date to A.D. 11. For discussion of these *leges*, see Robert E. Palmer (1974) 57-78, 240-244.

18 On the possible history of these prayers, see Harris 121, 265-266 with literature. For comparison of language with earlier sources, see Pighi 328-357 and Diehl 355-369.

19 On the *Fratres Arvales*, see Wissowa 561-564 and most recently Paladino, esp. 41-84. On the *carmen Arvale*, see Norden (1939) 109-280. See Liebeschütz 65 n.2 and Weinstock 217-220 for literature on prayers for the emperor. The standard edition of the *Acta Fratrum Arvalium* is by Henzen.

Among the literary sources for prayers, the earliest are the comedies of Plautus and Terence.[20] Although heavily influenced by the language and concepts of Greek models, Plautus, in particular, frequently employs authentic Roman religious language, especially brief colloquial petitions and emphatic oaths. In addition, Plautus' comedies contain several longer prayers. Although these are often parodies, or in fact precisely because they are parodies, they employ traditional structure and language. Otherwise, much of the humor would be lost. For example, in the *Persa,* there is a lengthy prayer of thanksgiving by the slave Toxilus, which draws much of its humor from parodying the traditional prayer of a victorious general (753-756).

A rare direct source for authentic prayers of private cult is the second century agricultural treatise of Cato the Censor, which describes in considerable detail the routine tasks of running a farm.[21] The thoroughly practical and mundane purpose of this treatise makes it quite valuable as a source of religious ritual and prayers. It includes several prayers to accompany typical sacrifices intended to ensure the prosperity of the estate: 1) the offering of wine to Jupiter Dapalis for the health of oxen prior to the spring sowing, 2) offerings of wine and sacrificial cakes to Janus and Jupiter, in combination with offerings to Ceres, prior to the harvest, 3) the sacrifice of a pig as a piacular offering prior to the thinning of a sacred grove, and 4) the procession and sacrifice of a pig, sheep, and bull (*suovetaurilia*) for the protection of fields, flocks, and all those who live or work on the farm, along with piacular sacrifices should the original obtain unfavorable or questionable omens (*Rust.* 132, 134, 139, 141). Although these prayers were intended for private use, their language is similar to that found in official prayers in later sources. Thus they are helpful in establishing a chronological framework for the use of particular religious terms and formulae in public, as well as private, prayers.

---

20 See Secknus' monograph devoted to the analysis of prayers in Plautus and Terence and their origins. See also Hanson *passim*. On a Plautine prayer of thanksgiving, see Fraenkel (1922) 236-240. On parodic prayers in Plautus, see Kleinknecht 157-178.

21 On the importance of Cato in the study of private cult, see De-Marchi 128-129. For texts of prayers with linguistic commentary, see Pisani 50-56. On their relationship to a similar prayer found in the Iguvine Tablets, see Berretoni 152-170. For a comparison of Catonian prayers and those of the *Ludi Saeculares*, see Diehl 356-369.

It is not until the first century that we encounter another author whose extant work includes authentic prayers. Cicero furnishes us with a variety of prayers belonging to official Roman cult. In the preface to his oration *pro Murena,* he recalls the official prayer to be offered at assemblies (*sollemnis comitiorum precatio*), which he himself had spoken on the occasion of Murena's election as consul. Other orations include proposals for decrees of supplications, which employ formulae identical to those spoken in the actual supplications (e.g. *Phil.* 14.37). References to such decrees, with the same language, also appear in orations and in letters (e.g. *Cat.* 3.15, *Fam.* 15.4.11). In his treatise on divination, other prayers and religious formulae occur (e.g. *Div.* 1.102).[22] Because of Cicero's position as a magistrate and augur, he is a credible source for such official prayers. His orations also, however, include more literary prayers. The oration against Verres, for example, concludes with one such prayer, sixty lines invoking thirteen deities by name. In addition, Cicero's works are a source for the brief colloquial prayers of daily life, such as *di omen avertant* (e.g. *Mur.* 88).

As a probable member of the college overseeing rituals of foreign origin, (*XV viri sacris faciundis*) and a friend of the augurs Messala and Cicero, Varro of Reate had excellent sources for the prayers which he preserved in his work on the Latin language, *de Lingua Latina,* and in his massive antiquarian study of *res divinae,* surviving only in quotations by later authors including Augustine.[23] Of particular interest for our study is the prayer which precedes summoning the people for the ritual close of a census, which Varro says is taken from the *censoriae tabulae* (*Ling.* 6.86). He also discusses the problems inherent in praying to unidentified spirits and the resulting formulae of invocation (*apud* Gell. *NA* 2.28.2-3). Finally, Varro is likely the ultimate source for much later scholarship on religious ritual and prayer which appears in Festus (via Verrius Flaccus), Macrobius, and Servius *auctus* (via Donatus).

It is four centuries after Varro that we may find our next significant source for authentic prayers. The *Saturnalia* of Macrobius includes two prayers that Scipio Aemilianus allegedly offered before his assault on

---

22 On the language of prayers and formulae in the *de Divinatione,* see the invaluable commentary of Pease 239-241.

23 On Varro's sources, see Cardauns 239-241 and Norden (1939) 4-5.

Carthage in 146 B.C. (3.9.7-11).[24] The *carmen evocationis* summons the tutelary deity of the city to abandon its people and places to destruction in return for a finer temple and worship in Rome. There follows a *carmen devotionis* of the lands and people of the enemy to Dispater, Veiovis, and the Manes as well as further prayers for the enemy's ruin. Macrobius attributes the preservation of these *carmina* to a certain Furius, perhaps Scipio's friend, L. Furius Philus, and gives as his own source the Severan antiquarian Serenus Sammonicus, who was himself perhaps dependent upon Verrius Flaccus.[25] While some scholars flatly deny the authenticity of these *carmina*, most consider them to be at least antiquarian reconstructions from genuine formulae.

The Servian commentaries on the poems of Vergil are a storehouse of information about Roman religious practices. These commentaries represent the work of two different editors, the late fourth century grammarian Servius and an anonymous antiquarian known as Servius *auctus* (or *Danielis*), who probably lived in the seventh century. Both men seem to have relied on a common source, the early fourth century commentary of Aelius Donatus, who in turn based his work in part on the learned writings of the Republican scholars Verrius Flaccus and Varro; thus the commentaries are a generally reliable source on Roman prayer.[26] On a number of occasions, they make reference to prayer formulae which Varro and Flaccus attribute to the priestly books. Also of great interest are discussions of the importance of hiding the real name of cities in order to avoid evocation by an enemy (Serv. *auct. ad Aen.* 2.351, 4.577).

---

24 On Macrobius, see Cameron 25-38. On the *carmen evocationis*, see Basanoff 17-30 and Köves-Zulauf, esp. 85-88, 100-104. On the *carmen devotionis*, see Köves-Zulauf 88-90; Versnel (1976) 365-410 with bibliography; cf. Janssen 357. Latte dismisses these *carmina* (82 n.4), but Rawson's careful discussion concludes that "we should probably . . . suspend judgement" as to their authenticity (173). On the language of these prayers, see Engelbrecht 478-484.

25 On Sammonicus, see Champlin 189-212. On the identification of Furius, see Rawson 163-164, 168-169.

26 On the sources of Servius and Servius *auctus*, see the recent discussions of Kaster 169-170 and Goold 102-117 with references.

I: The Predecessors of Livy and Vergil

Despite the different genres in which they wrote, Livy and Vergil shared a common perspective, which manifests itself in the authors' use of prayers. It is an epic perspective, which is revealed in the frequency of prayers, as well as in the general religious framework employed by both writers. The numerous prayers in both works reflect a similar world view, which characterizes the religion of the Roman state as well as heroic epic. For both, the world is inhabited by two classes of beings, the mortal and the immortal. Since earthly events directly relate to the attitudes and behaviors of divine beings, prayer is a common form of speech in both the literary world of epic and the real world of Roman life. This shared world view, however, does not completely explain the similarities between Livy's history and Vergil's *Aeneid.* An examination of references by Livy to the epic genre and a comparison of his use of prayer to that of other authors, both historiographers and epic poets, suggests that he consciously used prayers to contribute to the epic tone of his history. A comparison of the influences which the epics of Homer and Ennius had on prayers in the two Augustan works will also show how different choices of epic models produced differences in the language and structure of prayers in the later works.

Livy and Epic Poetry

In the preface, Livy himself comments on the affinity between parts of his history and epic poetry. Describing early portions of his work, he mentions *quae ante conditam condendamve urbem poeticis magis decora fabulis quam incorruptis rerum gestarum monumentis traduntur.* He continues, *datur haec venia antiquitati ut miscendo humana divinis primordia urbium augustiora faciat* (*praef.* 6-7). These words fit well with Servius' description of heroic epic as made up *ex divinis humanisque personis, continens vera cum fictis, . . .* (*ad Aen. praef.*). Given Livy's comments, it is appropriate that the early books of Livy's history show the heaviest poetic influence. This poetic influence is evident in diction and rhythm and also in the frequency of prayers; there are three times as many prayers in the first decade as in either of the two subsequent decades.[1] As McDonald comments, "where his subject matter lay in the legendary past—'poeticis decora fabulis'—Livy was ready to present his history as a prose epic, in the high poetic manner" (166).

Livy again points to the epic character of his history by concluding the preface with the approximation of an invocation. It is a common poetic device, in which the poet himself enters his created world and prefaces his literary endeavor with a prayer to its divine inhabitants that they favor his work. Homer and Vergil both open their epics with invocations to the Muses to aid them in their tasks, for example: μῆνιν ἄειδε, θεά, Πηληϊάδεω Ἀχιλῆος . . . (*Il.* 1.1) as compared with *Musa, mihi causas memora* . . . (*Aen.* 1.8).[2] In an allusion to such predecessors, Livy writes, *cum bonis potius ominibus votisque et precationibus deorum dearumque si, ut poetis, nobis quoque mos esset, libentius inciperemus, ut orsis tantum operis successus prosperos darent* (*praef.* 13=App. 1). Livy's diction here reflects language appropriate for an actual prayer (*dei deaeque*; *prosperos*; *darent*). By closing the preface with this pseudo-invocation, Livy alerts his audience that he views his history as similar in some ways to the epic poems of Homer, Ennius, and Vergil.

Livy's readers would have been familiar with a historian's use of certain poetical elements. In the *Institutiones*, Quintilian refers to the genre of history as *proxima poetis, et quodam modo carmen solutum est* (10.1.31). Livy's annalistic predecessors, especially Coelius and Antias, were also interested in the emotional effects characteristic of poetry and tragical history.[3] Furthermore, like Livy, his predecessors were particularly indebted to Ennius' *Annales*.[4] Shortly before the invocation, as Quintilian remarked not long after the history was published, Livy's words begin to move in the traditional epic meter, the dactylic

---

[1] On the much discussed poetic character of the first decade see *inter alios* Briscoe (1981) 13-15; Tränkle 103-149; Ogilvie 20-22; Walsh 248-249; Guittard (1981a) 33-44; McDonald 171; Gries 1-16; Stacey 17-82.

[2] On invocations, see Servius *ad Aen.* 1.8. No doubt Ennius also began his epic with a similar invocation; only a fragment remains: *Musae quae pedibus magnum pulsatis Olympus* (*Ann.* 1).

[3] On tragical history, see Walsh 40-42 and B.L. Ullman 25-53.

[4] This close relationship between historical epic and annalistic history partly grew out of common sources, particularly the priestly chronicles and national and family legends. Both also shared a common subject matter, *res gestae* of famous men. See Birch 88, Walsh 196-197, B. L. Ullman 62-69, and Häussler 114-118, 219-223 on the relationship between historical epic and historiography.

hexameter: *facturusne operae pretium sim* (*Inst.* 9.4.74). It is possible that this is an intentional allusion to Ennius.[5] Occasional poetic rhythms were not foreign to earlier historiographers; they appear in Coelius, Antias, and Sallust.

The epic character of certain portions of Livy's history has been noted before, but insufficient attention has been given to the contribution which prayers make to the epic quality of the work.[6] Prayers lend an epic color, not merely to the first few books of the history, but to the history as a whole. In his frequent inclusion of prayers, Livy departs from the practice of his historiographical predecessors. The poetical character of these prayers is shown by the contrast of Livy's history with earlier historiography and by certain similarities to poetic works, especially Ennius' *Annales*.

Prayers in Earlier Historiography

While considering the effect which Livy's prayers had on Augustan readers, it is important to remember that Livy's contemporaries would have read his history against the larger background of previous histories of Rome. A comparison of the frequency of prayers in Livy's work with the only other historiographical works which survive from the Republic, those of Caesar and Sallust, reveals a striking difference.

While 77 simple petitionary prayers appear in the surviving books of Livy, there is not a single such prayer in Sallust.[7] The closest approximation is a wish that the Numidian Adherbal addresses to the Roman Senate in his plea for aid against Jugurtha: *quod utinam . . . aliquando aut apud vos aut apud deos inmortalis rerum humanarum cura oriatur* (*Jug.* 14.21). One would expect at least some of Sallust's numerous

---

[5] On Livy's opening words, see Lundström 8-11. On the general use of metrical phrases in ancient historiography, see R. Ullman 62-69. Some scholars go to considerable lengths rearranging textual word order to produce metrical phrases; see for example, Morgan 61-66. The question of poetic sources, especially Ennius, for Livy's periodic dactylic rhythms has elicited much scholarly comment. Walsh takes a moderate approach, accepting some deliberate allusions to Ennius, but attributing other dactylic rhythms to Livy's personal prose form (137, 253-255). See Birch 88-94 for a more recent discussion of the topic with earlier bibliography.

[6] Guittard (1981a) discusses the contribution which *carmina devotionis* make to the epic color of the first decade (33-44).

[7] Livy's history also contains 24 vows and 27 oaths. See Appendix 4B.

speeches to contain divine petitions since they were a rhetorical commonplace, as we know from Servius' comment on the practice of invoking the gods at the beginning of a speech and as evidenced directly by Ciceronian speeches (*ad Aen.* 11.301; e.g. *Mur.* 1). Instead, Sallustian speeches never contain more than a brief exclamation, such as *per deos inmortalis* (e.g. *Cat.* 52.5), or an appeal for divine witnesses, e.g. *deos hominesque testamur* (*Cat.* 33.1).[8] Like his model Thucydides, Sallust displays little interest in religious matters. In fact, his characters on occasion express sceptical attitudes about the value of prayers. Speaking on the fate of the Catilinarian conspirators, Cato questions the usefulness of prayers in particular situations: *non votis neque suppliciis muliebribus auxilia deorum parantur; vigilando, agundo, bene consulendo prospere omnia cedunt. Ubi socordiae te atque ignaviae tradideris, nequiquam deos inplores: irati infestique sunt* (*Cat.* 52.29).[9]

Given the brevity that was probably characteristic of commentaries, Caesar's lack of prayers is less surprising. Only one brief petition appears, in indirect discourse: the eagle-bearer for the tenth legion begged the gods *ut ea res legioni feliciter eveniret* (*BGall.* 4.25.3). The difference between the world view presented in Livy's history and that in Caesar's commentaries is particularly clear when the reader considers their relative concern—or lack of concern—with religious matters in general and prayer in particular. Caesar describes a world where success or failure depends upon the actions of men and not of gods. Prayers have no function in such a world or in a literary work which records that world's history.[10]

In part, the difference between these historiographers and Livy is due to the annalistic tradition. When Livy devotes himself to religious matters, he is pursuing traditional topics of earlier annalistic histories. Like Livy's work, they included frequent references to vows, games, festivals, prodigies, and similar events. This emphasis derived from the

---

[8] Two pseudo-Sallustian letters, however, close with prayers (*Epist.* 1.8.10, 2.13.8). On the question of the authenticity of these letters, see Syme (1964) 315-351. On Sallust and religion, see Syme (1964) 246-247.

[9] Notice the prayer terminology *prospere . . . cedunt.* Cf. *Hist.* 1.77.3 *vatum carminibus* and 1.55.7 *Quirites . . . non prolatandum neque votis paranda auxilia.*

[10] Kajanto distinguishes between Livy's personal beliefs and the traditional piety of his characters (24-42). My own study is not concerned with what Livy personally believed. For an argument in support of Livy's religiosity, see Stübler *passim.*

priestly chronicles (*annales*), which determined the name and to some extent influenced the style and content of the annalistic tradition.[11] These priestly annals, however, were not literary works but bare chronological records of events.

Despite this traditional interest in religious matters, the virtual absence of prayers surviving from the early annalistic histories suggests that Livy's work was unusual in its high frequency of prayers. Only one simple petitionary prayer survives. Gnaeus Gellius' annals contain a prayer of Romulus' future wife, Hersilia, who appeals for divine intervention in the war between Sabines and Romans. She begins, *Neria Martis, te obsecro, pacem da, te uti liceat nuptiis propriis et prosperis uti* . . . (*apud* Gell. *NA* 13.23.13).[12] Certainly this was not the only prayer which appeared in earlier histories.[13] But the fact that only one remains among the fragments is telling, since many accounts of religious customs do survive. With its preservation of archaic language, prayers represent the type of material likely to be preserved by later grammarians. In contrast to the paucity of prayers in historiography, fragments from early poetry include a number of prayers.

Another indication of the differing use of prayers by Livy and his predecessors appears in the comparison of passages which handle the same material. For example, we are able to compare the way in which Claudius Quadrigarius and Livy tell the story of M. Valerius Corvinus' *cognomen*. Quadrigarius writes:

> Atque ibi vis quaedam divina fit: corvus repente inprovisus advolat et super galeam tribuni insistit atque inde in adversari os atque oculos pugnare incipit; insilibat, obturbabat et unguibus manum laniabat et prospectum alis arcebat atque, ubi satis saevierat, revolabat in galeam tribuni. Sic tribunus

[11] On the nature of the priestly chronicle and its relationship to the annalistic tradition, see Frier, esp. 270-275. Verbrugghe questions the widely accepted distinction of *annales* from other types of historiography (192-222).

[12] On this prayer and its relationship to Enn. *Ann.* 104 and 117, see Skutsch (1985) 246-247. Interestingly there is no parallel prayer in Livy for this event.

[13] This observation is made partially on the basis of the length of Gellius' work and also on the work of Dionysius, who cites Gellius six times and writes an equally full history including prayers, although fewer than Livy (e.g. the prayer of Aemilia: Dion. Hal. 2.68.4, a fetial oath: 3.3.5, a vow: 3.32.4). Plutarch, who was influenced by Dionysius, also includes prayers in his biographies. On Gellius, see Frier 310. On Livy's use of Gellius, see Luce 175 notes 84-85.

> spectante utroque exercitu et sua virtute nixus et opera alitis propugnatus, ducem hostium ferocissimum vicit interfecitque atque ob hanc causam cognomen habuit Corvinus (*apud* Gell. *NA* 9.11).

Compare Livy's version:

> Minus insigne certamen humanum numine interposito deorum factum; namque conserenti iam manum Romano corvus repente in galea consedit, in hostem versus. Quod primo ut augurium caelo missum laetus accepit tribunus, precatus deinde, si divus, si diva esset qui sibi praepetem misisset, volens propitius adesset. Dictu mirabile, tenuit non solum ales captam semel sedem sed, quotienscumque certamen initum est, levans se alis os oculosque hostis rostro et unguibus appetit (7.26.3-5=App. 29).[14]

The basic facts remain the same in the two authors. Both acknowledge the divine presence and identify the raven as a bird of omen (*ales*). What is distinctive about Livy's version is the inclusion of a prayer; it is brief, but complete and ritually proper for the occasion of an unsought omen. The tribune Valerius invokes no god by name because he is not certain which god has sent the omen. Instead he uses the appropriate technical formula for such a situation: *si divus, si diva*. The clause relative *qui . . . praepetem misisset* narrows the invocation to a specific, although unnamed, deity. Valerius petitions the deity's propitious aid with another technical formula: *volens propitius*.[15] Livy has described the same event as did his predecessor, but he has emphasized the religious element by including the human response to the divine manifestation by means of prayer. A similar pattern appears when comparing two passages from Valerius Antias and Livy, where the mere reference to a prayer in Antias contrasts with Livy's use of technical language appropriate for a prayer of thanksgiving (Antias *apud* Gell. *NA* 4.18.3; Liv. 38.51.7-11=App. 67).

---

[14] On these two passages, see Peter 210-212 and McDonald 158-159.

[15] Cf. Verg. *Aen.* 3.265-266=App. 86; Ov. *Met.* 9.701-703. The term "technical" refers to those words and formulae which appear repeatedly in actual prayers and perform a specific function in the regular structure and religious content of those prayers.

Another passage in which Livy apparently introduces a prayer is revealed by a comparison with Polybius, his primary source for the Hannibalic War. In Livy's history, Scipio includes a brief apotropaic prayer when he addresses the mutinous army at Sucro in 206 B.C.: *ne istuc Iuppiter optimus maximus sirit, urbem auspicato dis auctoribus in aeternum conditam huic fragili et mortali corpori aequalem esse* (28.28.11= App. 86). Polybius makes no mention of a prayer (11.28).

Poetic Influence on Prayers in Livy

In contrast to the relative paucity of prayers in historiography, prayers were a normal feature in poetry of many genres. As mentioned previously, the early comedies are replete with prayers, mostly brief colloquial exclamations. A few such appear in Livy, more likely derived from contemporary speech than from the comic poets. The fragments of early tragedies contain eighteen prayers.[16] But only one of the surviving prayers affords any evidence for their direct influence on Livy. The *Periboea* of Pacuvius contains a prayer whose construction and vocabulary find echoes in a speech of Scipio Africanus (*trag.* 297; Liv. 29.27.3= App. 55).

The surviving fragments of Ennius' *Annales* also include many prayers, which, unlike the dramatic fragments, do provide evidence of a significant influence on Livy.[17] Many scholars have considered the influence of the *Annales* on various aspects of Livy's history.[18] Although Livy only cites Ennius on one occasion, as he describes Fabius Cunctator

---

[16] Liv. Andr. 20-22; Pacuv. 112, 219-220, 295, 296-297, *praet.* 1; Naev. *BPunic* 17; Acc. *trag.* 52, 127, 164-168, 240-242, *praet.* 5-6, 15, 36; Enn. *scen.* 209, 284-286, 333, *inc.* 1. By the term "colloquial" I indicate that the language was part of everyday usage by the average Roman citizen and not restricted to a particular group of persons such as priests or magistrates, or to particular activities such as religious rituals.

[17] Prayers: V^3 52-53, 54, 104, 107-108, 110-114, 117, 208-210, 620=Skutsch 58-59, 26, 99, 102-103, 105-109, 100, 191-193, 619; curse: 360-361. See Skutsch (1985) 22-24 with references on Ennian influence of Livy.

[18] In addition to the references included in the following discussion, see Birch 88-94; Pinsent 13-18; Skutsch (1968) 51-53 and 88-92; Walsh 136-137; Aly *passim*; Norden (1915) 54-55 n.1, 114, 141 n.2, 155.

(Liv. 30.26.9),[19] there are many other passages where an Ennian influence is evident, although not credited by Livy.

Among the seven preserved Ennian prayers, that which bears the most striking similarity to one in Livy is an invocation of the Tiber River: *teque pater Tiberine tuo cum flumine sancto* (*Ann.* 54). This may have influenced the prayer Livy assigns to Horatius Cocles: *Tiberine pater, . . . te sancte precor, haec arma et hunc militem propitio flumine accipias* (2.10.11=App. 11). Notice the repetitions of *te, pater, Tiberine, flumine* and *sancte/o.*[20] If Livy did borrow from Ennius, he may have changed the context of the prayer in more significant ways than its wording, since a recent discussion of Ennius' prayer proposes that, like his heir Vergil, the earlier poet assigned these words not to Horatius, but to Aeneas.[21] Although Servius preserves a traditional invocation to the Tiber which may have influenced any of these three authors, the verbal similarities link Livy directly to Ennius, rather than to the traditional invocation, at least as found in Servius: *adesto, Tiberine, cum tuis undis* (*ad Aen.* 8.72-73).

Ennius' *Annales* includes another prayer which may have had a significant influence on various official petitions in Livy:

> Quod mihi reique fidei regno vobisque, Quirites,
> se fortunatim feliciter ac bene vortat (102-103 Skutsch).

It is the only example of an official usage of the petition *bene vortat* (q.v.) prior to Livy. All other examples occur in the comic authors, occasionally with a hint of parody of a formal prayer; otherwise, it seems to be a colloquial prayer. Yet Livy uses variations on this formula ten times in the history, always in official settings. The strong colloquial tradition of this petition, paired with its virtual disappearance outside of Livy, suggests that Livy departs from his usual practice of employing contemporary official formulae. A partial explanation for this stylistic departure is the desire for *variatio* with the similar, but contemporary, technical formula *quod bonum faustum felixque sit*. But it is noteworthy that Livy

---

[19] *Nihil certius est quam unum hominem nobis cunctando rem restituisse, sicut Ennius ait* (Liv. 30.26.9). Livy has transformed the original into *oratio obliqua*: *unus homo nobis cunctando restituit rem* (Enn. *Ann.* 370).

[20] On this Livian borrowing from Ennius, see Skutsch (1985) 186 and Walsh 256.

[21] See Jocelyn (1964) 293 and Norden (1915) 161-162. Vahlen CLIX assigns these lines to Ilia.

has chosen what may be an archaic prayer; he may be echoing consciously his predecessor's epic annals.

Another pair of texts from Ennius and Livy describes the response of the people to the apotheosis of Romulus. Cicero preserves the fragment from the *Annales*:

> . . . Iusto quidem rege cum est populus orbatus,
> "pectora diu tenet desiderium,"
> sicut ait Ennius, "post optimi regis obitum;
> . . . . simul inter
> sese sic memorant 'o Romule, Romule die,
> qualem te patriae custodem di genuerunt!
> o pater, o genitor, o sanguen dis oriundum!'"
> . . . (*hic dixit Cicero*),
> "'Tu produxisti nos intra luminis oras'"
> (*Rep.* 1.64).[22]

In Livy, the prayer to Romulus is cast in indirect speech:

> Romana pubes . . . tamen velut orbitatis metu icta maestum aliquamdiu silentium obtinuit. Deinde a paucis initio facto, deum deo natum, regem parentemque urbis Romanae salvere universi Romulum iubent; pacem precibus exposcunt, uti volens propitius suam semper sospitet progeniem (1.16.2-3=App. 2).

The passage in Livy likely reminded his readers of Ennius' words.[23] Both prayers hail Romulus as the offspring of a god. In Livy, the soldiers explicitly address Romulus as himself a god. The Ennian fragment too, by its resemblance to the beginning of a hymn with its invocation, genealogy, and aretalogy, suggests that the addressee is a god. Both prayers also further identify Romulus as the father of the Roman people. In both Cicero and Livy, the comparison of the people's grief at Romulus'

---

[22] On *diu* see Skutsch (1985) 257.

[23] On *salve parens* as an ancient formula of the Parentalia, the festival honoring ancestors, see Ogilvie (1965) 86. On *salvere* in Ovid *Fast.* 1.509, Bömer remarks "poetische Anrede . . . Dass die Anrede in Prosa vorkommt, wird durch Liv. 1.16.3 nicht wahrscheinlicher . . . Das bezieht sich auf die Apotheose des Romulus und kann zudem aus Ennius stammen." Skutsch (1968) comments "Romulus . . . is clearly not addressed as a god here . . . The passage . . . must have preceded the story of the deification." (132).

disappearance to that of orphaned children may recall such a comparison in Ennius (*orbatus, orbitatis*). Ennius' reference to their "grief" (*desiderium*) may also have influenced Livy's mention of it a few sentences later following the account of Proculus Iulius' vision of the deified Romulus: *desiderium Romuli . . . facta fide immortalitatis lenitum* (1.16.8). Finally both authors describe Romulus as a guardian of Rome. It is noteworthy that the prayer for Romulus to "protect his offspring" employs the rare verb *sospitare*. One of only four attestations of this verb occurs in Ennius, although not in this passage (*scen.* 295). It is possible that the verb appeared in the missing section of this text and thus inspired Livy's choice. Even if Livy is following a source other than Ennius for this prayer, his Augustan audience, who knew the *Annales* quite well, would probably have recalled the Ennian prayer when reading or hearing Livy's version.

Of course there are prayers in Ennius that do not appear in Livy and vice versa. For example, the prayer of Hersilia, wife of Romulus, which appears in Cn. Gellius' annals, may have an Ennian model (*apud* Gell. *NA* 13.23.13).[24] But in Livy's version of the abduction of the Sabine women, there is no mention of prayer to the gods.[25] Even when both authors include prayers in similar situations, those prayers do not always share similar language. Thus the ritual prayers preceding a *devotio* in the two authors bear no linguistic relationship to one another.[26]

Livy's more extensive use of prayers than his fellow historiographers is evident not only from passages in which he has introduced prayers not found in his sources, but also from the fact that only one petitionary prayer from an earlier annalist has survived.[27] Furthermore, Livy's history shows definite influence of Ennius' epic poem the

---

[24] On this prayer and its relationship to Enn. *Ann.* 104 and 117, see Skutsch (1968) 134-136.

[25] Cf. Liv. 1.11.2 *Romulum Hersilia coniunx precibus raptarum fatigata orat ut parentibus earum det veniam et in civitatem accipiat: ita rem coalescere concordia posse* and Liv. 1.13.2 where the Sabine women beg their husbands and relatives to cease war.

[26] *Divi hoc audite parumper, | ut pro Romano populo prognariter armis | certando, prudens animam de corpore mitto* (Enn. *Ann.* 208-210). Cf. Liv. 8.9.6-8=App.115.

[27] A similar dearth appears in non-annalistic sources; one brief vow remains: *Iuppiter, si tibi magis cordi est nos ea tibi dare potius quam Mezentio, uti nos victores facias* (Cato *Orig. apud* Macrob. *Sat.* 3.5.10).

Annales. Both authors make frequent use of prayers, some of which share similar contexts and wording. Because of these similarities and the frequency of prayers in other poetic sources, Livy's prayers lend a poetical tone to the history.

Ennian Influence on Prayers in the *Aeneid*

Ennius had a great influence on Vergil[28] as well as on Livy. Comparison of the two epic poems shows that Vergil adopted words, phrases, and even whole lines from his Roman predecessor although he always rearranged the order or changed at least one word. Ennius' influence, however, did not extend to the diction of prayers.[29] While Ennian prayers occasionally borrow the technical formulae and structures of official prayers, Vergil studiously avoids any impression of the technical language of Roman cult. The only exception is Aeneas' prayer to the Tiber after the river god has appeared to him in a nocturnal vision: *tuque, o Thybri tuo genitor cum flumine sancto* (*Aen.* 8.72=App. 136). Vergil is dependent here on the Ennian invocation discussed above as a model for Livy as well: *teque pater Tiberine tuo cum flumine sancto* (*Ann.* 54).[30] Although the meaning of the Vergilian prayer is very similar to the one in Ennius, there are significant differences in diction. Vergil has substituted the Greek name (*Thybris*) in the place of the traditional Roman epithet for the god of the Tiber River, *Tiberinus*, which Servius describes as the proper ritual terminology (*ad Aen.* 8.31).[31] He has also replaced the traditional term of respect (*pater*) with a poetic synonym (*genitor*).[32]

---

[28] The essential work on this topic is Norden (1915).

[29] This appearance may be due to the accidents of preservation, but a number of prayers from the *Annales* have been preserved. Seven prayers are included among the 628 verses or fragments of verses from the *Annales*.

[30] On the relationship between the invocations of Ennius and Vergil, see Skutsch (1985) 184-186, Lehr 41-42, and Norden (1915) 161-162.

[31] Servius comments: *in sacris Tiberinus, in coenolexia Tiberis, in poemate Thybris vocatur* (*ad Aen.* 8.31). Servius *auctus* writes: *Tiberinum . . . in hunc fluvium cecidisse et fluvio nomen dedisse*: *nam et a pontificibus indigitari solet* (*ad Aen.* 8.330). Thybris also occurs in Pallas' prayer (*Aen.* 10.421=App. 138). Vergil uses both names in non-prayer contexts.

[32] On *genitor* as a poetic gloss, see Cordier 159 and Lehr 34. Lehr follows Forcellini when saying incorrectly that *genitor* does not appear in Latin before Cicero (34). It is found in Ennius, *Ann.* 113; see Skutsch (1985) 256-259. On *pater* as the cultic term, see Lehr 34 and Skutsch (1985) 185-186. *Pater* is retained at 10.421. Note that Livy has employed the traditional noun and epithet.

Thus Vergil has lessened the official tone of the only prayer in the *Aeneid* for which a Latin model has been found.

Homeric Influence on Prayers in the *Aeneid*

In three instances, the affinity between the content and/or language of prayers of the *Aeneid* and those of the Homeric epics shows references to large sections of Greek prayers. Vergil follows his Greek model most closely in the prayer of the queen and women of Latium who approach the "temple and highest citadels of Pallas Athena":

> Nec non ad templum summasque ad Palladis arces
> subvehitur magna matrum regina caterva
> dona ferens, ...
> et maestas alto fundunt de limine voces:
> "armipotens, praeses belli, Tritonia virgo,
> frange manu telum Phrygii praedonis, et ipsum
> pronum sterne solo portisque effunde sub altis"
> (*Aen.* 11.477-479, 482-485=App. 106).

In the *Iliad*, Hecabe and the women of Troy also approach a "temple of Athena on the citadel of the city," where the priestess Theano leads them in prayer:

> Αἱ δ' ὅτε νηὸν ἵκανον 'Αθήνης ἐν πόλει ἄκρῃ,
> . . . .
> "πότνι' 'Αθηναίη, ῥυσίπτολι, δῖα θεάων,
> ἆξον δὴ ἔγχος Διομήδεος, ἠδὲ καὶ αὐτὸν
> πρηνέα δὸς πεσέειν Σκαιῶν προπάροιθε πυλάων"
> (6.297, 305-307).

Both passages represent the prayers of women seeking the preservation of their cities against a powerful enemy. They ask Athena to break the spear of that enemy and to lay him face down before the gates of their cities. In both cases, Athena denies the women's prayers.

One of the most interesting situations for prayer in both authors is the hero's encounter with a goddess disguised as a mortal. Odysseus and Aeneas find themselves in places they do not recognize (Ithaca and Carthage) and they seek its identity from apparent strangers (Athena and Venus). Aeneas addresses Venus thus:

> nulla tuarum audita mihi neque visa sororum,
> o quam te memorem, virgo? namque haud tibi vultus
> mortalis, nec vox hominem sonat; o, dea certe
> (an Phoebi soror? an Nympharum sanguinis una?),
> sis felix nostrumque leves, quaecumque, laborem
> et quo sub caelo tandem, quibus orbis in oris
> iactemur doceas: ignari hominumque locorumque
> erramus vento huc vastis et fluctibus acti.
> multa tibi ante aras nostra cadet hostia dextra
> (*Aen.* 1.326-334=App. 134).

Odysseus greets Athena:

> Ὦ φίλ', ἐπεί σε πρῶτα κιχάνω τῷδ' ἐνὶ χώρῳ,
> χαῖρέ τε καὶ μή μοί τι κακῷ νόῳ ἀντιβολήσαις,
> ἀλλὰ σάω μὲν ταῦτα, σάω δ' ἐμέ· σοὶ γὰρ ἐγώ γε
> εὔχομαι ὥς τε θεῷ καί σευ φίλα γούναθ' ἱκάνω.
> καί μοι τοῦτ' ἀγόρευσον ἐτήτυμον, ὄφρ' ἐῢ εἰδῶ·
> τίς γῆ, τίς δῆμος, τίνες ἀνέρες ἐγγεγάασιν;
> ἦ πού τις νήσων εὐδείελος ἦέ τις ἀκτὴ
> κεῖθ' ἁλὶ κεκλιμένη ἐριβώλακος ἠπείροιο;
> (*Od.* 13.228-235).

Both Aeneas and Odysseus address the strangers as divinities and ask them to be propitious. Then both ask to what land or to what country they have come.

Aeneas' prayer to Venus also recalls that of Telemachus to Odysseus, whom Athena has disguised as a beggar. After marveling at the change, Telemachus says:

> Ἦ μάλα τις θεός ἐσσι, τοὶ οὐρανὸν εὐρὺν ἔχουσιν·
> ἀλλ' ἵληθ', ἵνα τοι κεχαρισμένα δώομεν ἱρὰ
> ἠδὲ χρύσεα δῶρα, τετυγμένα· φείδεο δ' ἡμέων
> (*Od.* 16.183-185).

With its identification of the stranger as a divinity and the direct request for him to be propitious, Telemachus' greeting resembles that of Aeneas to Venus more closely than does that of Odysseus to Athena. Telemachus, unlike Odysseus, but like Aeneas, promises sacrifices. Finally, Telemachus' prayer, like that of Aeneas, includes the ironic element that the person not recognized is a parent.

There is also a similarity between two prayers for success in battle. In the *Aeneid*, Arruns prays to Soractean Apollo that he may strike down the female warrior Camilla:

> Summe deum, sancti custos Soractis Apollo,
> quem primi colimus, cui pineus ardor acervo
> pascitur, et medium freti pietate per ignem
> cultores multa premimus vestigia pruna,
> da, pater, hoc nostris aboleri dedecus armis,
> omnipotens. . . . .
> . . . haec dira meo dum vulnere pestis
> pulsa cadat, patrias remeabo inglorius urbes
> (*Aen.* 11.785-790, 792-793=App. 108).

In the *Iliad*, Achilles prays to Dodonean Zeus as he sends Patroclus into battle:

> Ζεῦ ἄνα, Δωδωναῖε, Πελασγικέ, τηλόθι ναίων,
> Δωδώνης μεδέων δυσχειμέρου· ἀμφὶ δὲ Σελλοὶ
> σοὶ ναίουσ' ὑποφῆται ἀνιπτόποδες χαμαιεῦναι.
> ἠμὲν δή ποτ' ἐμὸν ἔπος ἔκλυες εὐξαμένοιο,
> τίμησας μὲν ἐμέ, μέγα δ' ἴψαο λαὸν 'Αχαιῶν,
> ἠδ' ἔτι καὶ νῦν μοι τόδ' ἐπικρήηνον ἐέλδωρ·
> αὐτὸς μὲν γὰρ ἐγὼ μενέω νηῶν ἐν ἀγῶνι,
> ἀλλ' ἕταρον πέμπω πολέσιν μετὰ Μυρμιδόνεσσι
> μάρνασθαι· τῷ κῦδος ἅμα πρόες, εὐρύοπα Ζεῦ,
> . . . .
> αὐτὰρ ἐπεί κ' ἀπὸ ναῦφι μάχην ἐνοπήν τε δίηται,
> ἀσκηθής μοι ἔπειτα θοὰς ἐπὶ νῆας ἵκοιτο
> τεύχεσί τε ξὺν πᾶσι καὶ ἀγχεμάχοις ἑτάροισιν
> (*Il.* 16.233-241, 246-248).

Vergil uses Homer's prayer to suggest general content rather than specific details. Most obviously, both Arruns and Achilles seek to check the disheartening success of an enemy warrior and pray that the one accomplishing that escape unharmed. While Achilles addresses Zeus as worshipped in the ancient cult of Dodona, Arruns directs his prayer to the god of another time-honored cult: that of Soractean Apollo. Knauer notes that both of these cults are distinguished by practices involving the feet.[33]

---

[33] On these two prayers, see Knauer 310-314. Other parallels between the two prayers show "die 'Treue' der Nachbildung in der vollkommenen 'Umkehrung' der

Most striking is the similarity of the divine response to the two prayers. In response to Arruns' prayer:

> Audiit et voti Phoebus succedere partem
> mente dedit, partem volucris dispersit in auras
> (*Aen.* 11.794-795).

In response to the prayer of Achilles:

> Τῷ δ' ἕτερον μὲν δῶκε πατήρ, ἕτερον δ' ἀνένευσε·
> νηῶν μέν οἱ ἀπώσασθαι πόλεμόν τε μάχην τε
> δῶκε, σόον δ' ἀνένευσε μάχης ἒξ ἀπονέεσθαι
> (*Il.* 16.250-252).

These are the only two passages where either poet describes a response in which the god answers one part of a prayer, but denies another portion.

Considering the extent of Ennian influence on the *Aeneid* as a whole, it is remarkable that Vergil has borrowed so little from the Latin poet in the wording of prayers. This may represent a conscious decision to avoid the official tone of some Ennian prayers. Instead, in at least three prayers, Vergil has looked to his Greek predecessor for models.

Vorlage besteht." While Achilles prays for glory and the return of his armor, Arruns disavows any desire for glory or the armor of his enemy (311).

II: The Invocation

The invocation is that part of a prayer that attempts to gain the attention of a deity, a necessity for any petition to be effective.[1] Not all prayers contain invocations; many abbreviated prayers make no reference at all to their divine addressees, for example, *quod bonum faustum felixque sit.* Every complete formal prayer, however, regardless of purpose (e.g. vow, propitiation), does include an invocation. Its usual location is the beginning of a prayer, preceding any specific request. The simplest invocations contain only the names of the divinities addressed. When known, an added epithet may specify a particular aspect of the deity, for example Jupiter Optimus Maximus. Sometimes honorific epithets, such as *pater*, or persuasive ones, such as *omnipotens,* are included.

Correct identification of the deity to be invoked was of the utmost importance; there would be absolutely no value in making a request of the wrong god. Varro illustrates the absurdity of such an attempt with a reference to the comic playwrights: *optemus a Libero aquam, a Lymphis vinum* (*apud* Aug. *De Civ. D.* 4.22). Proper identification, however, was no simple task. In book 8 of the *Aeneid*, Evander describes one aspect of the problem:

> "Hoc nemus, hunc," inquit, "frondoso vertice collem
> (quis deus incertum est) habitat deus" . . . (8.351-352).

It was relatively easy to recognize the presence of a divinity in those numinous woods, but not so easy to know which particular divinity was manifested. Varro discusses a similar uncertainty about addressing the divinity responsible for an earthquake: *quoniam, et qua vi et per quem deorum dearumve terra tremeret, incertum esset* (*apud* Gell. *NA* 2.28.3). Recognition of divine activity was only the first step in phrasing a proper invocation. In the case of tutelary deities of cities, the problem of identification was complicated by the intentional hiding of names. According to the religious understanding of the early Romans and other peoples, the dangerous power of evocation possessed by anyone knowing

[1] On the *invocatio* and its formulae, see Appel, names: 75-94, epithets: 94-109, 114. See also Ogilvie (1969) 24-29.

the real name of a city's tutelary deity made it essential to hide that name.[2] In addition to all of these problems, there was the uncertainty caused by the multiple aspects, and thus epithets, which an individual deity might possess. For example, the Romans variously invoked Jupiter as Optimus Maximus, Feretrius, and Stator, among other names. Therefore, it was possible to know part of a divinity's name and yet still be unable to utter a proper invocation.

One way in which the Romans dealt with the problems of phrasing a proper, and thus effective, invocation was to include various formulae which might compensate for possible errors.[3] When nothing more was known than that some divine force was present, the generic formula *si divus, si diva* substituted for the unknown name and avoided excluding or offending the deity. This formula was especially connected with the evocation of the tutelary divinity of a foreign city. When the uncertainty focused on the correctness of the name, a formula, such as *sive quo alio nomine te appellari volueris* (Serv. *auct. ad Aen.* 4.577), might avoid divine displeasure. According to Servius *auctus*, the Roman priests used this phrase even when addressing the supreme deity of their own state.[4] Another means of compensating for uncertainty about a divine name was to identify the deity's residence or sphere of influence as, for example, in the clause *cuius in tutela urbs est.*[5] Even when naming a specific deity, the petitioner might take precautions against a mistaken name or the omission of a name. In these situations, the petitioner both addressed the specific deity and included a universal invocation such as *dique omnes* (e.g. Liv. 1.32.9=App. 143).[6] It is this practice that Plautus parodies in

[2] On the secret name and evocation, see Pliny *HN* 28.18; Macrob. *Sat.* 3.9.1-8; Serv. *auct. ad Aen.* 2.351; Basanoff 25-30; Köves-Zulauf 68-69 n.13, 85-102 with literature.

[3] On the use of various relative clauses in Greek and Latin prayers and hymns, see Norden (1913) 143-145 and 168-176. On these clauses as a possible origin of epic epithets, see Versnel (1981) 14.

[4] Cf. *sive vos quo alio nomine fas est nominare* (Macrob. *Sat.* 3.9.10). In Macrobius the reason for the petitioner's uncertainty is the foreign setting.

[5] On this formula, see below s.v.

[6] Commenting on this formula in Vergil, Servius writes: *post specialem invocationem transit ad generalitatem, ne quod numen praetereat, more pontificum*; Servius *auctus* adds: (per) *quos ritu veteri in omnibus sacris post speciales deos, quos ad ipsum sacrum, quod fiebat, necesse erat invocari, generaliter omnia numina invocabantur* (*ad Geor.* 1.21). Cf. Serv. *ad Aen.* 8.103. See also the comments of Zwierlein on Catullus' reference to this formula in 66.9 (276-277).

the oath of Alcesimarchus, who begins by calling to witness *di deaeque, superi atque inferi et medioxumi* . . . (*Cist.* 512). After a bungled attempt to name specific deities with their appropriate genealogies, he concludes by returning to another modified universal invocation: *di . . . omnes, magni minutique et etiam patellarii* . . . (*Cist.* 522).

In addition to the deity's name along with modifiers, invocations sometimes include a verbal request for the deity to direct attention to the speaker. Although such requests are generally confined to poetic prayers, the fetial oaths of Livy's history consistently ask that a deity "listen" (*audi*). This request may tell us more about the language of oaths than of prayers in general and thus will be discussed in the chapter on oaths. Besides requesting that the gods listen, poetic prayers may also ask the deity to "approach" (*adi*) or "look at" (*aspice*) the speaker. These requests are common in Greek literary prayers, where they may have originated.

Language of Livy and Vergil
I. Verbs of Invocation

adi[7]

Aen. 8.302

Although *adire* does on occasion refer to the real or figurative approach of gods to men (Plaut.. *Curc.* 262; Ov. *A. A.* 2.496), Vergil is the only author to use this verb in a prayer.[8] It occurs in one of two prayers containing the phrase *pede secundo*: to Hercules, *et nos et tua dexter adi pede sacra secundo* (8.302) and to Cybele, *adsis pede, diva, secundo* (10.255). While *adi* is similar in meaning ("come") to *adsis*, which appears in the parallel prayer, there may be a distinction between the two verbs based on the idea of epiphany in *adire*, as opposed to assistance in *adesse* (q.v.). In the Salian hymn to Hercules, the priests are repeating a ceremonial invocation for the divinity's presence at a feast in his honor. In contrast, the prayer to Cybele, using *adsis*, seeks her aid in battle.

---

[7] Discussions of individual words and phrases include all examples which appear in authentic prayers of the Republic or the Augustan period. Examples of literary prayers are selected on the basis of similarity to the usage in Livy and Vergil. Citations of examples show the range of authors in which a particular usage appears and thus refer to at least one early and one first-century author if such exist.

[8] For additional examples of the approach of gods to men, see TLL 2.626.8-29. Cf. similar Greek expressions ἐλθέ (*Il.* 23.770) and ἱκοῦ (Eurip. *Orest.* 1231); see Ausfeld 516-517.

aspice
Aen. 2.690

Only poetic prayers include a petition that the gods "look" at the petitioner's situation as an antecedent to a more situation-specific request.[9] The verb *aspicere* first occurs as such a preface in Catullus' lamentation, *o dii . . . me miserum aspicite* (76.17-19). There is only one Augustan attestation. Before requesting that Jupiter confirm the omen of flame above Iulus' head, Anchises asks for the god's attention: *aspice nos* (Verg. *Aen.* 2.690=App. 82). Commenting on this passage, Servius writes, *quia intuentes dii iuvant: unde est* (10.473) *atque oculos Rutulorum reicit arvis, et contra* (1.482) *diva solo fixos oculos aversa tenebat.* The anthropomorphic nature of the request that a god "look" is emphasized in Iarbas' angry speech to Jupiter concerning the relationship of Dido and Aeneas: *aspicis haec*? (4.208) and in Jupiter's response: *audiit Omnipotens, oculosque ad moenia torsit* (4.220). Unlike the former passage (2.690), Iarbas' prayer does not request that the deity attend to the speaker and any other request is only implied.[10]

II. Descriptive Clauses

qui . . . colis
Liv. 5.21.3, 24.38.8, 29.27.2

The semantic usage of *colere* varies from the literal notion of "inhabit" to the more abstract concepts of "cherish," "protect," "be the guardian of" (*Lewis and Short*). Outside of Livy, there are no prose prayers that contain the verb *colere* in relative clauses identifying a deity, but there are numerous poetic prayers which do.[11] For example, Plautus' *Poenulus* includes a prayer offered by the Carthaginian Hanno to the resident gods of the foreign city Calydon: *deos deasque veneror qui hanc urbem colunt* (*Poen.* 950). The use of such clauses may be traced through Greek literature as far back as Achilles' prayer to Apollo in the beginning of the *Iliad*:

[9] This practice is also characteristic of Greek poetic prayers (e.g. ἴδοι: Aesch. *Suppl.* 206). For additional examples, see Ausfeld 516.

[10] For additional examples of *aspicere* and related verbs, see Appel 118 and TLL 2.836.72-78.

[11] For examples, see Appel 110-111.

κλῦθί μευ, ἀργυρότοξ', ὃς Χρύσην ἀμφιβέβηκας
Κίλλαν τε ζαθέην Τενέδοιό τε ἶφι ἀνάσσεις (37-38).[12]

A comparison of examples from Livy with other prayers supports the thesis that his models are literary. In Camillus' *evocatio* of the tutelary deity of Veii, the general invokes Juno *regina, quae nunc Veios colis* (5.21.3). If Livy's clause is a literary variation, perhaps the technical formula occurs in the comparable *carmen evocationis* from Macrobius, which employs the phrase *cui populus civitasque Carthaginiensis est in tutela* (*Sat.* 3.9.7).[13] A comparative model exists for another such clause in Livy. Before assaulting the city of Henna, Pinarius invokes the resident divinities: *di qui hanc urbem, hos sacratos lacus lucosque colitis* (24.38.8=App. 47). The related verb *incolitis* appears in a Ciceronian invocation of the divine residents of the same city: (*sc. Ceres et Libera*) *quae illos Hennensis lacus lucosque incolitis* (*Verr.* 5.188). The similarity between the two prayers is underscored by the unusual phrase *lacus lucosque.*[14] Cicero's prayer is a highly rhetorical one, invoking numerous divinities, each distinguished by a different descriptive clause; it does not resemble any of the actual prayers that have survived from official cult. Livy consistently follows the contextual paradigm observed in Plautus; in all three prayers, the speakers are ignorant of the correct name of the deities whom they address. The reason is always the same; they are gods of a foreign territory. Camillus, Pinarius, and in a third prayer, Scipio Africanus (29.27.2=App. 55) anticipate military encounters with an enemy and thus an unknown tutelary deity.

cuius in tutela est
Liv. 34.24.2

There are a significant number of references that connect the phrase *in tutela est* with one particular type of prayer, the *carmen evocationis.*[15] One of the two extant versions of that prayer, found in Macrobius, invokes the tutelary deity of Carthage thus: *si deus, si dea est, cui populus*

---

[12] On the literary development of relative clauses identifying divine power, see Norden (1913) 168-176. See also Ausfeld 524-525.

[13] Stübler notes the possibility of an Ennian model for this prayer (50 n.20).

[14] Cf. *lucus locusve* (*Act. Frat. Arv. a.* 183 Henzen CLXXXVI). *Lacus* and *lucos* appear in a list of sacred places in Lucret. 5.75. See Appel 114 for additional examples of *incolere*, including Liv. 6.16.2.

[15] Cf. the usage of *tutela* as a legal technical term; see VIR 5.1146-1153, esp. 1151.24-29.

civitasque Carthaginiensis est in tutela (*Sat.* 3.9.7). A similar phrase occurs twice in Macrobius' preceding discussion of the ceremony as well (3.9.2, 3). It also appears in a votive inscription naming the commander of the siege of Isaura Vetus in 75 B.C.: *sei deus seive deast, quoius in tutela oppidum vetus Isaura fuit* (*AE* [1977] 237, no. 816). This association is further supported by references to the ceremony by two scholars. Citing Verrius Flaccus, Pliny writes of the customary practice in sieges of evoking the god *cuius in tutela id oppidum esset,* and thus the importance of hiding the name of Rome's own protecting deity *in cuius dei tutela Roma esset* (*HN* 28.18). Similarly, Servius *auctus* writes of hiding *in cuius dei tutela urbs Roma sit* (*ad Aen.* 2.351). Not all occurrences of this phrase appear in association with evocations. The acts of the Arval Brotherhood for A.D. 183 refer to sacrifices to be made to several named deities and lastly to *sive deo sive deae, in cuius tutela hic lucus locusve est* (Henzen CLXXXVI). These various examples show that the phrase regularly occurs in combination with the indefinite invocation *si deus, si dea,* as a means of compensating for the vagueness of that address.

Livy's usage, however, does not conform to this pattern. His version of the *carmen evocationis,* as mentioned above, contains the clause *quae nunc Veios colis* (5.21.3=App. 21). He uses the phrase *in tutela est* instead on a different occasion, a brief informal prayer by the Achaean praetor Aristaenus for the safety of Argos: *ne istuc . . . Iuppiter optimus maximus sirit Iunoque regina, cuius in tutela Argi sunt, ut . . .* (34.24.2=App. 60). Here the clause seems to explain the motivation for divine action. It certainly does not identify unnamed deities.

qui . . . praesidet
Liv. 31.51.8
Aen. 3.35

There are relatively few Latin prayers that contain relative clauses with the verb *praesidere* to indicate a place of divine protection. The usage of these clauses, however, is strikingly consistent. Of five examples in Ciceronian orations, four invoke the tutelary divinities of Rome (*Verr.* 5.188 refers to Sicilian deities). For example, the following prayer comes from Cicero's speech on the Manilian law: *testorque omnis deos et eos maxime qui huic loco temploque praesident* (70; cf. *Sull.* 86, *Dom.* 144, *Phil.* 13.20). The association of the verb *praesidere* with the tutelary deities of Rome is further supported by two later texts that use the words *praesidium* and *praeses* in connection with the divinities of

Rome. In a prayer to the deified Julius Caesar, Valerius Maximus writes: *oro ut propitio ac faventi numine tantorum casus virorum sub tui exempli praesidio ac tutela delitescere patiaris* (1.6.13; n.b. the collocation of *praesidio* and *tutela*). Servius *auctus* makes a relevant comment on the usage of the term *praeses* to describe the Capitoline goddess Minerva: *qui "praeses" . . . (legunt) quasi quae praesideat rei (intellegunt). Sic praesides dii urbis appellantur* (*ad Aen.* 11.483).

In Livy, the one prayer containing the clause under discussion also refers to tutelary deities. Scipio Africanus expresses his intention to honor the anniversary of his Carthaginian victory: *ad Iovem optimum maximum Iunonemque et Minervam ceterosque deos qui Capitolio atque arci praesident salutandos ibo* (38.51.8=App. 67). If *praesidere* is a technical term, as seems likely, Livy's words may echo the no-longer extant formulae of official thanksgiving ceremonies addressed to the Capitoline deities.

Vergil's *Aeneid* also contains one prayer in which the invocation includes the formula *qui . . . praesidet* to identify the deity addressed. In response to the omen of blood dripping from the roots of shrubs growing over the body of Polydorus, who was foully murdered in Thrace, Aeneas invokes Mars, *Geticis qui praesidet arvis* (3.35). This usage is consistent with that of the comparative prose examples, which use the clause to identify tutelary deities.

quicumque
Aen. 1.330, 9.209

Sometimes *quicumque* appears in a literary variant for the technical formulae *vel quo alio nomine te appellari volueris* or *sive quo alio nomine fas est nominare* (Serv. *auct. ad Aen.* 4.577; Macrob. *Sat.* 3.9.10). Thus, using a phrase which approximates the official formula, Horace addresses a jar of Massican wine in a prayer parody: *quocumque lectum nomine Massicum* (*Carm.* 3.21.5).[16] Livy uses *quicumque* when he is uncertain exactly how he should refer to Aeneas: (*sc. tumulus Aeneae*) *situs est, quemcumque eum dici ius fasque est, super Numicum flumen: Iovem indigetem appellant* (1.2.6). Although this passage is not a prayer, its phrasing, particularly *fas . . . est*, echoes the invocation formula found

---

[16] See Norden (1913) 143-147 on this Horatian usage. Cf. Catull. 34.21. For other examples, see Appel 78-79.

in Macrobius. In addition, the verb *dici* recalls *appellari* and *nominare* of the two versions of the formula.

The pronoun *quicumque* appears in two prayers of the *Aeneid.* In one Vergilian passage, the adjective *quaecumque* clearly functions as a substitute for an unknown divine name:

> . . . O dea certe
> (an Phoebi soror? an Nympharum sanguinis una?)
> sis felix nostrumque leves, quaecumque, laborem
> (1.328-330=App. 134).

The pronoun also underscores the irony of the situation in which Aeneas is unable to recognize his own mother Venus, disguised as a Tyrian maiden. Another prayer containing *quicumque* may also have an ironic flavor. Using an oath, Nisus affirms his belief in Euryalus' bravery:

> . . . non ita me referat tibi magnus ovantem
> Iuppiter aut quicumque oculis haec aspicit aequis
> (9.208-209=App. 171).

From the perspective of religious language, the clause beginning *aut quicumque* could refer to the possibility that the speaker is invoking the god by an incomplete or incorrect name. On the other hand, the words seem to call into question Jupiter's fairness (*oculis . . . aequis*) and suggest that some other unknown deity may be a more appropriate addressee. In fact, neither Nisus nor Euryalus return from their mission to summon Aeneas; perhaps Nisus did call upon the wrong god after all.

quisquis
Liv. 3.25.8
Aen. 4.577, 9.22

Servius *auctus* compares the technical formula *sive quo alio nomine te appellari volueris* to the poetic phrase *quisquis es* in the prayer Aeneas voices after Mercury has warned him to leave Carthage:[17]

> . . . sequimur te, sancte deorum,
> quisquis es, . . . (4.576-577=App. 89).[18]

---

[17] See also Serv. *auct. ad Aen.* 2.351 for a comparison of the pontifical formula and *quisquis es* with reference to the secret name of Rome.

[18] Cf. similar prayers in Ov. *Met.* 3.611-614 and Tib. 2.2.2. For additional examples in Augustan poetry, see Appel 78-79.

Servius notes that there are multiple aspects of Mercury, which could make Aeneas question the correct name to use in invoking the god. Alternatively, Aeneas may be uncertain about the identity of the god who has sent Mercury as his messenger, just as in the prayer Turnus offers after the appearance of Iris (cf. Serv. *ad Aen.* 9.16, 22):

. . . sequor omina tanta,
quisquis in arma vocas . . . (9.21-22).

Clearly the expression *quisquis* here does not indicate that Turnus is uncertain about Iris' identity. Turnus recognized the goddess and even addressed her by name. In both prayers, the concern of the petitioner remains the same, to address the deity correctly. Although not a variant of the pontifical formula, *quisquis* serves a similar function by compensating for the speaker's ignorance of a deity's correct name. It is possibly modeled on the Greek phrase ὅστις ποτ' ἐστίν, which addresses the same concern.[19] Already in Plautus the phrase appears in close relation to a Latin prayer. Palaestra prays to the unknown deity before whose shrine she has washed ashore: *quisquis est deus, veneror ut nos ex hac aerumna eximat, . . .* (*Rud.* 257).

In Livy, the clause *et quidquid deorum est* also compensates for uncertainty about the appropriate deity to invoke.[20] Rebuffed by the Aequian general, a Roman envoy thus asks the gods to witness treaty violations: *sacrata quercus* (*sc. Iuppiter*) *et quidquid deorum est audiant* (3.25.8=App. 148). The envoy is in a foreign state and may refer specifically to his ignorance of local deities. In addition, use of the coordinating conjunction *et* following a named addressee recalls the universal invocation *et ceteri di*, which regularly compensated for the speaker's neglect of some deity.

si divus si diva
Liv. 7.26.4

The technical formula *si divus si diva* and its variants occur in a variety of sources that address unknown divinities.[21] It first appears in the

---

[19] On this Greek phrase, see the comments of Martinez 54-55, Fraenkel (1950) on Aesch. *Ag.* 160, and Norden (1913) 145-146 n.3.

[20] Once in reference to an oath, the phrase simply substitutes for naming the specific gods: *iurantes per quidquid deorum est* (Liv. 23.9.3).

[21] On the concept of unknown deities, see Versnel (1981) 15; Wissowa 37-38. For a commentary on the various witnesses to this formula, see Alvar (1985). See also the

prayer that Cato prescribes to expiate the pruning of a sacred grove: *si deus, si dea es, quoium illud sacrum est* (*Rust.* 139). Because the speaker does not know the name of the minor divinity or genius protecting the grove and thus, not the gender either, the formula substitutes the choice of *deus* or *dea* and a descriptive relative clause. By including the possibility of either gender, the supplicant ensures that the divinity will not ignore the petition on account of improper identification. This is the same situation that Varro describes with reference to the offering of sacrifice and prayer following an earthquake: *si deo, si deae* (*apud* Gell. *NA* 2.28.2). There is also epigraphical evidence for the formula; three brief dedicatory inscriptions from the Republic address an otherwise unidentified deity (CIL 1.801, 1.1485, 1.2644=ILLRP 291-293). The *Acta Fratres Arvales* provide evidence for the continued or renewed use of this formula at least into the third century of this era (Henzen CLXXXVI).[22]

The situation most clearly connected with this formula is the ritual of *evocatio* which preceded assault on a foreign city. Because of the divinity's great power to grant or to remove protection of a city, the divine name was kept secret from all but those priests who needed to know it for performance of their ritual duties. Thus, at the siege of Carthage in 146, Scipio Aemilianus could only invoke that city's deity by means of a paraphrase: *si deus, si dea est cui populus civitasque Carthaginiensis est in tutela* (Macrob. *Sat.* 3.9.7). In another discussion of the ceremony of evocation, Servius *auctus* notes that a shield dedicated to the protecting deity of Rome bore a similar paraphrase: *genio urbis Romae sive mas sive femina* (*ad Aen.* 2.351; cf. *sim.* Laev. *frg.* 26) Further confirmation of the use of this phrase in evocations may be provided by a recently found inscription recording the fulfillment of a vow to the tutelary deity of Isaura Vetus, which was captured by Servilius Vatia in 75 B.C:

> Serveilius . . . sei deus seive deast, quoius in tutela oppidum vetus Isaura fuit, [. . .] votum soluit.[23]

---

discussion of this formula and its origins in Greek prayer style in Martinez 54 and Norden (1913) 143-146.

[22] It is uncertain whether Arnobius' reference to the formula reflects contemporary usage (*Adv. Nat.* 3.8).

[23] In addition to Alvar (1985) 247-248 on the formula, see Hall 568-571; *AE* (1977) 237 no. 816; Le Gall 519-524.

The one example of this formula in Livy conforms to the pattern of unknown deities. After receiving an omen prior to individual combat, M. Valerius invokes the responsible god: *si divus si diva esset qui sibi praepetem misisset* (7.26.4=App. 29).[24] The omen manifests the attention of an individual deity, whose identity, however, Valerius is unable to ascertain. Most other Livian prayers are addressed purposefully to a specific deity or to the *dii immortales*, situations in which appropriate epithets or restrictive phrases are available to clarify the identification.

Vergil's *Aeneid* includes one variant on this technical formula. As in the example from Livy's history, this passage also describes the response to a portent. The harpy Celaeno has just prophesied the troubles awaiting the Trojans in Italy. The men respond in fear:

> . . . cecidere animi, nec iam amplius armis,
> sed votis precibusque iubent exposcere pacem,
> sive deae seu sint dirae obscenaeque volucres (3.260-262=App. 86).

Vergil has altered the formula semantically; the alternative here is between goddesses and ill-omened birds rather than god and goddess. He has maintained the sound of the original formula by his choice and placement of the word *dirae*.

---

[24] If this anecdote records an actual event, it would date the use of this formula to the mid-fourth century. See Alvar (1985) 253 on the possibilities. See also Alvar (1985) 254 for the association of this event with a possible *evocatio*.

III: Simple Petitions

This chapter examines the language of simple petitionary prayers, excluding the sub-types of vow and oath, each of which is discussed in succeeding chapters. The function and occasions of petitionary prayers are summarized in the prologue of this book.

The Language of Livy and Vergil
I. Verbs of Petition

In a formal prayer, one or more verbs of petition (*verba precandi*) signal the speaker's intention to make a request; such verbs appearing in Livy and Vergil are *adoro, oro, precor, quaeso, veneror*, and *veniam peto*. Traditionally they stand near the beginning of the prayer, often directly following the invocation and preceding the requests.[1]

Although by the time of their appearance in literature of the third century, all but one of these verbs (*adoro*) introduce petitions in prayer, and thus in this context may be translated "request," they differ in semantic history and etymology. Although the first four verbs are common in secular language, *veneror* and *veniam peto* are closely tied to the religious sphere; the former is used only in a religious sense during the Republic. In addition, the first four refer only to the act of the speaker in making his request, but *veneror* and *veniam peto* also refer to the good will of the deity invoked.[2] Such a usage illustrates well the Roman concept that the gods must be favorably disposed toward the speaker before they will fulfill any request. The subordination of the notion of favor to the point where the phrase *veniam peto* is simply one of a number of *verba precandi*, in fact underscores the fundamental character of this notion. It is likely that the authors using these technical terms were not

---

[1] On *verba precandi*, see Guittard (1980) 395-403.

[2] See Szantyr 26-33, 43-44. Schilling (1954), based on his interpretation of the root *ven* as "charm," views the differences between these verbs as even more radical in origin; it is the difference between juridical and *venusienne* religion (52-59). He suggests that the reverential connotation of these words probably reflects a movement away from the original magical perspective. As a translation of *veneror ut* at this magical stage, he offers: "j'use de charme religieux pour obtenir" (36).

conscious of, or at least not particularly interested in, the concept of propitiation inherent in them.

adoro

Aen. 10.677

It is not until the first century B.C., in the poetry of Laevius, that *adorare* is attested as a *verbum precandi*:

> Venerem igitur almum adorans
> sive . . . (*frg.* 26).

As here, *adorare* generally serves to describe or introduce a prayer in the narrative.[3] Although it does not occur within the context of any prayer in prose, *adorare* does appear occasionally in poetic prayers—first in Vergil. Thus, in the midst of his petition that the winds halt the ship onto which Juno has lured him, Turnus interjects, *volens vos Turnus adoro* (10.677=App. 104). The only other Augustan example occurs in Propertius' prayer for Cynthia's continuing love: *maneat sic semper, adoro . . .* (1.4.27).

oro

Liv. 8.9.7

Oro first appears as a *verbum precandi* in Plautus. The old man Megaronides jokingly prays that Callicles' wife survive him: *deosque oro ut vitae tuae superstes suppetat* (*Trin.* 57).[4] Approximately three centuries later, Valerius Maximus offers the following prayer to the deified Julius Caesar: *tuas aras tuaque sanctissima templa, dive Iuli, veneratus oro ut propitio ac faventi numine tantorum casus virorum sub tui exempli praesidio ac tutela delitescere patiaris* (1.6.13). In addition to these literary uses, *oro* also appears in the official prayers of the Severan *Ludi Saeculares*, and in all likelihood, of the Augustan games as well: *precamur o[r]amus obsecramusque* (CIL 6.32329.13=Pighi 157.13).

In Livy's account of the *devotio* of Decius Mus, a unique series of verbs introduces the petition that the gods favor the Roman people: *precor venerorque veniam peto oroque, uti populo Romano vim victoriam*

---

[3] E.g. Verg. *Aen.* 1.48; Liv. 6.12.7; for additional examples, see Appel 65 and TLL 1.819.9-37. On Livian usage of *adorare*, see Gries 19-20. Cf. use in narrative (Liv. 7.39.13).

[4] Also in narrative in *Epid.* 302, *Merc.* 908, and *Poen.* 1134. Wagenvoort (1980) does not recognize a genuine use in prayer before Vergil (202 n.18); *contra* Szantyr 3. See TLL 9.2.1046.17-41 for additional examples.

prosperetis (8.9.7=App. 115; o. *Forchhammer*: feroque *mss*). The evidence of similar verbal combinations in numerous prayers indicates that this series should consist of *verba precandi*.[5] Therefore, *feroque* of the manuscript tradition should be emended to *oroque* as Madvig, Weissenborn, and others have done following Forchhammer.[6]

In the one example from the *Aeneid*, Vergil does not use *oro* to introduce a clause, but instead employs a one word accusative object, *vitam*. Evander pleads with the gods to protect the life of his son Pallas and to make his own life conditional upon that of his son:

> at vos, o superi, et divum tu maxime rector
> Iuppiter, Arcadii, quaeso, miserescite regis
> et patrias audite preces: si numina vestra
> incolumem Pallanta mihi, si fata reservant,
> si visurus eum vivo et venturus in unum:
> vitam oro . . . (8.572-577=App. 99).

precor
Liv. 2.10.11, 5.21.3, 24.38.8, 37.36.6
Aen. 4.621, 10.461, 12.179, 12.777

With the exception of the comic playwrights, literary sources, including Livy and Vergil, choose the single verb *precor* more often than any other *verbum precandi*.[7] The earliest usage of the verb *precor* to introduce a petition occurs in Ennius:

> te sancta precor, Venus, te genetrix patris nostri,
> ut me de caelo visas cognata parumper
> (*Ann*. 52-53; *trad*. saneneta, *emend*. sancta *Lindsay*).

In actual prayers a synonym such as *quaeso*, *veneror*, or *veniam peto* usually accompanies *precor*. In the private prayers that Cato recommends for use on the farm, however, the verb *precor* appears without a verbal synonym: *bonas preces precor uti* . . . (*Rust*. 134.2-3; 139). Here the *figura etymologica* strengthens the sense of the verb.

---

[5] See examples in Szantyr (7).

[6] In support of the transmitted reading *fero*, Wagenvoort (1964), following a dynamistic interpretation, writes *quod manifestam reddit mea quidem sententia mutuam veniae vim et naturam* (51 n.1).

[7] Plautus does use *precor* once with a religious sense (*Rud*. 640). On the secular origin of *precor*, see Schilling (1954) 54; *contra*: Wagenvoort (1964) 52.

Livy's preference for the simple *precor* over the official formula *precor quaesoque* does not consistently reflect the formality of the context in which the prayers occur. One of the four prayers containing the single verb *precor* does appear in an informal setting: Horatius' plunge into the Tiber River—certainly an occasion for brevity (2.10.11=App. 9). Although another petition is spoken by the commander Scipio Africanus to an envoy of King Antiochus, the conversation is a private one and the petition is merely an interjection: *aliis, deos precor, ne umquam fortuna egeat mea* (37.36.6=App. 65). Another prayer occurs at the close of Pinarius' exhortation to his troops before assaulting Henna:

> vos, Ceres mater ac Proserpina, precor, ceteri superi inferni-que di, qui hanc urbem, hos sacratos lacus lucosque colitis, ut ita nobis volentes propitii adsitis, si vitandae, non inferendae fraudis causa hoc consilii capimus (24.38.8=App. 47).

It is a formal prayer of moderate length with a full invocation, in other words, a likely context for the use of the official formula instead of the single verb. Finally, the *carmen evocationis* that Camillus speaks before attacking Veii would seem an ideal setting for official language, but Livy again eschews the traditional formula in favor of the simple *precor*. In fact, the *carmen* as a whole is a relatively brief prayer lacking a particularly official tone: *te simul, Iuno regina, quae nunc Veios colis, precor, ut nos victores in nostram tuamque mox futuram urbem sequare, ubi te dignum amplitudine tua templum accipiat* (5.21.3=App. 21; cf. Macrob. *Sat.* 3.9.7-8). Livy's literary tone here suggested to Stübler the possibility of Ennian influence (50 n.20).[8]

Of prayers in the *Aeneid* which use a *verbum precandi*, all but one employ *precor* (*quaeso* appears in 8.573). Two of these prayers are spoken before combat in petitions for divine aid (10.461, 12.777=App. 103, 110). A third is a brief parenthetical request in the middle of Aeneas' oath solemnizing his treaty with the Latins:

> et pater omnipotens et tu Saturnia coniunx
> (iam melior, iam, diva, precor, tuque inclute Mavors) . . .
> (12.178-179=App. 173).

[8] Alvar (1985) also questions this prayer's avoidance of the technical formula *sive deus sive dea* (257-258).

In addition to these propitiatory petitions, Dido concludes her curse of Aeneas with the words, *haec precor* (4.621=App. 90; cf. Audollent 274).

precor quaesoque
Liv. 9.8.8, 29.27.2

Clearly the formula *precor quaesoque*, with minor variations, is the preferred technical formula for use in the State cult from the first century. When Cicero invokes the Capitoline triad in his oration *de Domo Sua*, he varies the formula with the conjunction *atque* (144.6). In the prayers for the Augustan *Ludi Saceulares*, the order is inverted (CIL 6.32323.93=Pighi 114.93). The prayers of the *Fratres Arvales* naturally employ the plural (Henzen LVII; CXL). Cato's earlier prayer for the lustration of a farm shows use of the formula in a private setting: *Mars pater, te precor quaesoque* (*Rust.* 141). Livy uses the formula *precor quaesoque* to introduce two prayers: an emotional appeal at the close of Sp. Postumius' speech to the Senate concerning the Caudine Peace (9.8.8=App. 34) and the lengthy and highly formulaic prayer of Scipio Africanus before his expedition to Africa (29.27.2=App. 55).

With the exception of two secular passages in Livy, all attestations of this formula occur in religious contexts. Thus Livy gains a very dramatic effect from his use of *precor quaesoque* in the emotional appeal of Pacuvius Calavius for his son not to assassinate Hannibal and in the appeal of Sophoniba for Massinissa not to surrender her to the Romans (23.9.2, 30.12.13). The traditional religious usage underscores the gravity of these situations since the human/divine relationship is also one of powerless to powerful. The gods, like Massinissa and the son of Pacuvius Calavius, hold in their hands the power of life and death.

quaeso
Aen. 8.573

Among Latin authors, only Plautus and Terence prefer *quaeso* over *precor*, perhaps because of its frequent colloquial usage (e.g. Plaut. *Rud.* 499; Ter. *Ad.* 275).[9] Although Vergil also prefers *precor*, he does use its synonym in one prayer, that of Evander as his son departs for war:

[9] In his discussion of the secular character of *quaeso* and *precor*, Schilling (1954) points out Plautus' avoidance of the latter, but does not mention his use of the former (54). See Benveniste 2.158-161 for a recent discussion of the two verbs.

At vos, o superi, et divum tu maxime rector
Iuppiter, Arcadii, quaeso, miserescite regis . . .
(8.572-573=App. 99).

veneror
Liv. 8.9.7

Unlike the verbs previously discussed, *veneror* was limited to religious contexts throughout the Republic.[10] Although its etymology is much debated, its function in prayers is clear; it appears as a *verbum precandi* as early as Plautus, where it occurs in four prayers. For example, the Carthaginian Hanno, who is searching for his daughters and nephew, begins his prayer: *deos deasque veneror . . . ut . . .* (*Poen.* 950-951; see also *Bacch.* 173, *Rud.* 257, 1349). The verb probably has petitionary force also in a fragmentary prayer from Ennius' *Annales*, from which only the invocation remains and no petition, *Quirine pater veneror Horamque Quirini* (117). In Macrobius' version of the *carmen evocationis*, it is one of three synonyms introducing the petition: *precor venerorque veniamque a vobis peto ut . . .* (*Sat.* 3.9.7). This verb appears in only one prayer in Livy, the *devotio* of Decius Mus, where it belongs to a unique series of *verba precandi* very similar to that found in Macrobius' prayer of evocation: *precor veneror veniam peto oroque uti . . .* (8.9.7).

veniam peto
Liv. 7.40.4, 8.9.7

Concerning the noun *venia*, Servius writes, *proprie verbum pontificale est* (*ad Aen.* 1.519). Its use in religious contexts to indicate divine favor is attested from the time of Plautus (*Rud.* 26-27).[11] When combined with a verb such as *peto*, a formula of petition results. Thus, at the beginning of his oration *pro Rabirio*, Cicero introduces a petition to the gods with the following words: *pacem ac veniam peto precorque ab eis ut . . .* (*Rab. Perd.* 5).[12] In the *carmen evocationis*, Macrobius uses the phrase next to the related word *veneror*: *precor venerorque veniamque a vobis peto ut . . .* (*Sat.* 3.9.7). A very similar series introduces the

---

[10] On the religious character and use of *veneror*, see Szantyr's discussion with literature 26-33, 43-44.

[11] On the etymology of *venia*, see especially Szantyr 26, 32; Morani 44-47; Wagenvoort (1964) 57-58; Schilling (1954) 39-42.

[12] Cf. earlier example in Pacuv. *trag.* 296, . . . *veniam precor petens, ut . . .* On the collocation of *pax* and *venia*, see Schilling (1954) 41.

petitions that appear in Livy's account of the *devotio* of Decius Mus: *vos precor veneror veniam peto oroque, uti* . . . (8.9.7=App. 115). Livy uses a different *verbum precandi* (*poscere*) to describe Corvus' request for the glory of reconciling the mutinous soldiers at Capua: *adoravi veniamque supplex poposci ut* . . . (7.40.4=App. 31). As with *veneror*, these interchangeable uses of a request for divine favor and simple verbs of petition show how fundamentally connected the two ideas were.

The petition for *venia* acquired the additional notion of seeking pardon for an offense which might cause a deity to be angry and thus not favorably disposed toward the speaker's request (e.g. Verg. *Geor.* 4.536; Ov. *Met.* 4.632-633; *Fast.* 4.755).[13] Thus before speaking of the Bacchic rites, Hispala seeks divine indulgence: *pacem veniamque precata deorum dearumque, si coacta caritate eius silenda enuntiasset* (Liv. 39.10.5= App. 68). Similarly, Livy reports that in order to avert a plague in 291, the Senate ordered the people *supplicatum ire pacemque exposcere deum . . . Matres . . . veniam irarum caelestium finemque pesti exposcunt* (3.7.7-8).

Although *venia* does not appear in prayers in Vergil's *Aeneid*, it does appear in religious contexts (e.g. *Aen.* 3.144, 4.50).

II. Requests

Obviously, the most important element of a petitionary prayer is the actual request identifying the behavior or action desired from the deity. It is this for which other elements of the prayer prepare. For the sake of discussion, these requests may be organized into two broad categories: A) those which focus on the attitude of the gods and B) those which focus on the earthly event. This division is, of course, purely abstract; a favorable attitude on the part of the gods generally results in favorable circumstances on earth.

Petitions normally follow the *verba precandi*, often in a dependent clause. Sometimes, especially in poetry, the *verba precandi* and *preces* are arranged paratactically. The verbs of these independent clauses occur in either the subjunctive or the imperative.

---

[13] See Non. 185. For additional examples, see Appel 119-122. The Roman idea of pardon, unlike the Christian, was not one of an ethical sin.

A. Divine Attitude

Roman prayer, public and private, speaks plainly of a concern that the gods be favorably disposed toward human beings. Such an attitude is, according to Roman ideas, the determining factor in the granting of prayer. Thus, prayers commonly include a general request that the gods be well-disposed toward the speaker and the specific petition. These requests identify various related aspects of a favorable disposition, including approval, consent, will, and a more general favor. In addition to these concerns found in both prose and poetry, poetic prayers sometimes request divine pity.[14]

Divine Approval and Consent

adnuere

Aen. 9.625

Vergil is the first extant author to use *adnuere* ("to approve") in a prayer. The prayer *audacibus annue coeptis,* which had already appeared in the *Georgics* (1.40), reappears as Ascanius' petition to Jupiter before shooting Remulus (*Aen.* 9.625=App. 137). Other Augustan and post-Augustan poets also employ *adnuere* in prayers requesting divine favor toward some undertaking (e.g. Tib. 2.5.121; Ov. *Am.* 3.2.56).[15]

approbare

Liv. 10.13.12, 36.7.21

Although the verb *approbare* ("to approve") takes *deus* as its subject as early as Plautus (*Poen.* 1255), its usage in prayer is first attested in Cicero. It is with a tone of irony that Cicero prays for divine approval of Verres' actions: *at enim idcirco navem Mamertinis non imperasti, quod sunt foederati. Di adprobent! Habemus hominem in fetialium manibus educatum, unum praeter ceteros in publicis religionibus foederum sanctum ac diligentem* (*Verr.* 5.49). Cicero uses the colloquial formula without ironic intent when writing about his daughter's marriage (*Att.* 6.6.1, *Fam.* 2.15.2, 2.18.3).[16]

---

[14] Servius describes *venia* as a *verbum pontificale* and writes: *dicta autem venia ad eliciendam misericordiam* (*ad Aen.* 1.519). See the discussion under *veniam peto.*

[15] For additional examples, see Appel 138 and TLL 1.790.30-49. Note the derivative *numen* and its application to divine will and power.

[16] See also Ps. Sall. *Epist.* 1.8.10. Cf. *dis adprobantibus* (*Fam.* 10.22.1, *sim.* 1.9.19) and *auguriis adprobantibus* (Fest. 380L.=297M.).

In its two occurrences in Livy, the tone of the exclamation *di approbent* varies. When Fabius says *dei approbent . . . quod agitis acturique estis* to the people who have just illegally re-elected him to a second consulship within ten years, this plea for divine approval may ironically express his own disapproval (10.13.12=App. 38). But when Hannibal concludes his speech to Antiochus with *dii approbent eam sententiam quae tibi optima visa fuerit,* he does not know that the king will neglect his advice concerning war with Rome; he genuinely seeks divine blessing on the king's choice (36.7.21=App. 64). Only the audience, not the speaker Hannibal, recognizes the ironic tone of the formula.

fas est
Liv. 1.16.7, 1.18.9
Aen. 6.63, 6.266

In a lengthy monograph on the word *fas,* Cipriano distinguishes a number of different nuances of this technical term, two of which apply to prayers in Livy and Vergil.[17] In situations where there is no established standard for "what is lawful" as a result of divine assent, it is necessary to determine whether a particular action meets with divine approval.[18] One way to make this determination is simply by asking through prayer. Thus, seeing a vision of the deified Romulus, Proculus Iulius prays, *ut contra intueri fas esset* (Liv. 1.16.7=App. 3). The same phrase is used by the augur inaugurating Numa: *si est fas hunc Numam Pompilium . . . regem Romae esse* (Liv. 1.18.9=App. 5). In the case of the *inauguratio,* the god is expected to communicate what is lawful by means of auspices. So too, Cicero uses the technical phrases, *fas est* and *nefas est,* when referring to *auspicia* that would prohibit assemblies (*Dom.* 39, *Vatin.* 20, *Div.* 2.42). A similar prayer for divine permission is frequently used by poets who are about to describe something divine. Vergil seeks such permission before disclosing the secrets of the underworld:

> sit mihi fas audita loqui; sit numine vestro
> pandere res alta terra et caligine mersas
> (*Aen.* 6.266-267=App. 94).[19]

---

[17] On the related adjective *fastus,* see Michels 52-54 and Cipriano 96-107.

[18] Cipriano 54-56. I do not agree with all of her identifications but find her categories helpful.

[19] See Cipriano 56 on this passage. Cf. Catull. 51.2; Verg. *Geor.* 1.125-127; Hor. *Carm.* 2.19.9-11. For similar examples, see TLL 6.288.51-63.

In other situations, *fas est* should be understood in the context of the *ius divinum* (Cipriano 40). Unlike the previously discussed prayers in which the deity may give or withhold assent, in these cases an assumption is made that a particular act meets with divine approval. Of special concern was the invocation of a deity by the proper name. The term *fas* may have belonged to the traditional formula of invocation addressed to a god whose name was unknown or uncertain: *sive quo alio nomine fas est nominare* (Macrob. *Sat.* 3.9.10). In two passages, Livy paraphrases this invocation formula. First, he wonders what the proper name for Aeneas is: *situs est, quemcumque eum dici ius fasque est, super Numicum flumen: Iovem indigetem appellant* (1.2.6).[20] Second, Livy describes the term *rex* as a proper epithet for Jupiter: *quo Iovem appellari fas sit* (3.39.4).

Another way in which Livy uses the term *fas* to refer to the *ius divinum* is by its personification in the fetial formula.[21] After invoking Jupiter and the territory of the offending nation, the formula for satisfaction of wrongs concludes with the invocation, *audiat Fas* (1.32.6=App. 142). In a passage referring to fetial ritual, T. Manlius opens a speech concerning Latin requests with the invocation *audi Iuppiter haec scelera . . . audite Ius Fasque* (8.5.8=App. 153).

sinere
Liv. 1.32.8, 4.2.8, 28.28.11, 34.24.2
Aen. 9.409

The technical usage of the verb *sinere* ("to allow") in early Latin prayers is confirmed by its appearance in the archaic *carmen Arvale*: *neve luerve Marmar sins incurrere in pleores* (Henzen CCIV.32-37= CIL6.2104.32-37).[22] It also appears in Cato's prayer for the lustration of a field: *utique tu fruges, frumenta, vineta virgultaque grandire beneque evenire siris* (*Rust.* 141.3; cf. *di immortales adprobent beneque evenire sinant*: Ps. Sall. *Epist.* 1.8.10). Actual prayers from later periods, with the possible exception of Macrobius' *carmen devotionis*, afford no

---

[20] A discussion of the word *ius* goes beyond the scope of this study. See Dumezil 41-45 and Fugier 128-132.

[21] See Cipriano 74-78 on the "vincolo sacrale." On possible Greek literary influence, see Ogilvie (1965) 130; *contra*, Dumezil (1970) 91-92.

[22] On *sinere* in prayer, see Norden (1939) 130-133. For its use as an augural term, see Linderski (1986a) 2285.

additional examples of *sinere* introducing a positive request.[23] Poetry, however, does. Such a request occurs in the Plautine prayer, offered by the Carthaginian Hanno:

> . . . ut gnatas et mei fratris filium
> reperire me siritis . . . (*Poen.* 952-953).

Augustan poets also include such prayers occasionally, including Vergil. Before attempting to rescue the captured Euryalus, Nisus prays, *hunc sine me turbare globum* (9.409; cf. Ov. *Fast.* 4.914; Prop. 1.19.16).[24]

The more frequent context for *sinere* in prayers is a prohibitive clause. Thus, Livy uses it in the concluding self-curse of the fetial formula for redress of wrongs: *si ego iniuste impieque illos homines illasque res dedier mihi exposco, tum patriae compotem me numquam siris esse* (1.32.7=App. 142).[25] A common petition, which appears in Livy, is the apotropaic exclamation *ne (deus) sirit* (q.v.).

velle
Liv. 9.8.10
Aen. 1.733

The necessary pre-condition for the granting of any prayer request is the willingness of the gods.[26] Two of Plautus' characters give lively expression to this idea:

> LY: . . . deos credo voluisse; nam ni vellent, non fieret, scio.
> EUC: At ego deos credo voluisse ut apud me te in nervo enicem (*Aul.* 742-743).

The importance attached to divine willingness is demonstrated by the frequency of the official formula *volens propitius sis*. Occasionally in Augustan literature, the finite verb *velle*, which does not appear in any

---

[23] Büchler emends *siritis* to *servetis* at *Sat.* 3.9.11.

[24] When discussing the meaning of *sine* in *Aen.* 9.409=App. 100, Conington-Nettleship wonders whether "the word was suggested by anything more than metrical convenience."

[25] Ogilvie (1965) points to the perfect subjunctive in the fetial oath as proof of its late composition (131).

[26] Cf. *dis benevolentibus* (Plaut. *Mil.* 1351), *dis volentibus* (e.g. Plaut. *Pers.* 332; Liv. 37.19.5), *te volente* (Catull. 61.64, 69, 74), *volente deo* (Verg. *Aen.* 1.303), and "God willing." Cf. also *sei dei volent* (e.g. Plaut. *Poen.* 910). Secknus thinks this expression is modeled after the greek ἄν οἱ θεοὶ θέλωσιν (36). Cf. also Servius *auctus ad Aen.* 3.457: *volens quia cogi non potest si nolit, <aut> ut in sacrificiis 'uti volens propitiusque.'*

non-literary prayers, serves as a variant for the technical formula (Ov. *Met.* 7.37, 15.633, *A. A.* 3.347). At the beginning of her banquet for the Trojans, Dido requests an atmosphere of conviviality thus:[27]

> Hunc laetum Tyriisque diem Troiaque profectis
> esse velis . . . (*Aen.* 1.732-733=App. 79).

Livy uses the verb only once in a prayer. At the close of his speech concerning the Caudine Peace, Sp. Postumius prays to the gods, *precor quaesoque . . . novos consules legionesque Romanas ita cum Samnite gerere bellum velitis, ut omnia ante nos consules bella gesta sunt* (9.8.8-10=App. 34). The combination of *velle* plus an infinitive allows the author to conflate the petitions for divine favor and a situation-specific action.

Divine Favor

bonus

Aen. 1.734, 12.176, 647

As early as Terence, the adjective *bonus* occurs as an attribute of divinities in the colloquial exclamation *di boni.*[28] In the *Aeneid, bonus* occurs three times with the meaning "propitious," all in prayers and in reference to gods. Dido invokes the presence of *bona* Juno at the banquet for her Trojan guests (1.734=App. 79). According to Servius, one possible interpretation of *bona* here is *propitia, id est non irata Troianis, ut sis bonus o felixque tuis* (*ref. ad Ecl.* 5.65).[29] Dido's choice of words seems ironic since Juno was the goddess whose wrath had caused the Trojans' troubles. Her words may also seem ironic to the audience which knows Dido's own fate at the hands of this deity. Vergil again applies the adjective to Juno in a parenthetical petition within the oath that Aeneas swears to Latinus. This time there is no hidden irony; the comparative (*melior*) emphasizes the difference between the goddess' past emnity and her desired favor:

---

[27] Servius comments: *Esse velis: secundum Etruscam disciplinam locutus est; sic enim dicunt "volens propitius sis"* (*ad l.* 1.733). See also *ad* 3.457.

[28] For additional examples, see TLL 2.2086.31-43. See Secknus (46) on possible origin of this expression from the Greek ὦ φίλοι θεοί (Menander *frg.* 429). Cf. the ironic use of *bonus* in curses (e.g. Plaut. *Casin.* 238).

[29] Servius continues: *aut magis "bona" caelestis; est enim et inferna, ut Iunoni infernae dictus sacer.*

esto nunc Sol testis et haec mihi Terra vocanti,
quam propter tantos potui perferre labores,
et pater omnipotens et tu Saturnia coniunx,
(iam melior, iam, diva, precor) (*Aen.* 12.176-179=App. 173).

The usage of *bonus* in Turnus' prayer to the gods of the dead (*vos o mihi, Manes, este boni*) evokes various interpretations (12.646-647=App. 109). Vergil may be simply using a standard epithet of the gods. Alternatively, he is perhaps playing on the meaning of Manes ("propitious gods, be propitious").[30] Or Vergil is making an ironic comparison of the possible attitude of the supposedly malevolent souls of the dead toward Turnus with that of the traditionally benevolent gods of the upperworld.

dexter
Aen. 8.302

Vergil is the first Latin author to use *dexter* with the meaning "propitious" concerning a divinity. In their prayer to the Greek deity Hercules, the Salian priests request the god's propitious presence: *et nos et tua dexter adi pede sacra secundo* (*Aen.* 8.302=App. 97). Vergil also uses the adjective in a non-prayer context to describe the goddess *Fortuna* (*Aen.* 2.388). On both occasions, the Servian commentaries explain this novel usage of *dexter* as meaning "propitious." Commenting on *fortuna dextra*, Servius *auctus* writes, *favens*: *propitia ut "laeva" contraria.* Other authors, primarily poets, followed Vergil's example. Ovid, for example, addresses the prayer *dexter ades* once to Germanicus, as though he were divine, and twice to Janus (*Fast.* 1.6, 67, 69). In a variation on *volens ac propitius* (q.v.), Quintilian invokes the emperor Domitian as a god, *dexterque ac volens adsit* (*Inst.* 4. *praef.* 5). An older and continuing religious usage of *dexter* describes favorable auspices. Thus, Paulus Festus writes, *dextera auspicia*: *prospera* (65L.=74M.).[31]

---

[30] Festus writes: *ut Inferi di Manes pro boni dicantur a suppliciter eos venerantibus propter metum mortis* (132L.=146M.). Cf. CE 1164: *Di Manes, manes sitis.*

[31] For examples of *dexter* to describe auspices, see TLL 5.924.12-18; 61-84. Cf. *Aen.* 4.578-579=App. 89 *sidera . . . dextra.* But cf. *sinistra meliora auspicia quam dextra* (Fest. 454L.=339M.). On this problem, see Pease on Cicero's words: *nobis sinistra videntur, Graiis et barbaris dextra meliora* (*Div.* 2.82) and on *Div.* 1.12 *a laeva.* See also Linderski (1986b) 339-340.

felix

Aen. 1.330

Vergil is the first extant author to use *felix* ("propitious") as a divine attribute.[32] No other author uses the prayer *sis felix*. It is perhaps a Vergilian substitution for the technical but unmetrical *volens ac propitius sis*. The only prayer in the *Aeneid* that uses *felix* is the request addressed by Aeneas to the unrecognized Venus: *sis felix* (1.330=App. 134). A similar prayer appears in the *Eclogues*; the deified Daphnis is addressed *sis bonus o felixque tuis*! (*Ecl.* 5.65). It is possible that this apparently novel usage is attributable to the ambiguous status of the addressees in these two prayers. Although Aeneas professes to believe that the maiden he has encountered must be a goddess, the innovative usage of this adjective may suggest some uncertainty on his part. Daphnis' status as a deified human being is also ambiguous. In the language of authentic prayers, the only attested use of the adjective *felix* occurs in the formula *quod bonum faustum felixque sit* and its variants. Servius comments: *sis felix propitia. felix enim dicitur et qui habet felicitatem et qui facit esse felicem* (*ad Aen.* 1.330).

placidus

Aen. 3.266, 4.578

Placidus does not appear in prose prayers at all and is not employed in poetic prayers until the Augustan period.[33] In fact, the adjective is not used of gods until the Augustan period. When attributed to gods, the primary meaning, "tranquil," acquires the additional connotation of "propitious." Two Vergilian prayers provide the first examples of this usage. In response to the harpy Celaeno's frightening prophecy, Anchises prays that the gods may avert the omen and *placidi servate pios* (*Aen.* 3.266=App. 86; *var.* placide *Pgamma*[1]). Aeneas, too, prays that a god be *placidus* when he responds to Mercury's admonition that he depart from Carthage: *adsis o placidusque iuves* (4.578=App. 89). Here, however, the idea of "tranquil" may persist since Aeneas continues his prayer with a

---

[32] There is a prayer in Silius that uses *felix* of an immortal: *nympha . . . nobis felix oblata secundes* (Sil. *Pun.* 8.227-228). For additional examples of *felix* as a divine attribute, see TLL 6.439.16-30.

[33] A sepulchral monument does show the adjective applied to the *manes*: *Vos ite placidi* (CIL 5.3653).

request for good, i.e. placid, weather.[34] Horace and Ovid both offer near contemporary examples of similar prayers that employ *placidus* to request that a god listen propitiously to requests:

condito mitis placidusque telo
supplices audi pueros, Apollo; . . .
(Hor. *Car. Saec.* 33)

huc ades atque audi placidus, Neptune, precantem
(Ov. *Met.* 8.598).[35]

In addition to a new meaning for *placidus*, Vergil's usage also displays a novel construction with the verbs *servare* and *iuvare* (q.v.), which regularly stand alone or, in the case of *iuvare*, with the adverb *bene*.[36]

propitius
Liv. 39.11.7

In addition to its usage in the official formula *volens propitius sit*, the adjective *propitius* appears alone in combination with the verb *esse*. Plautine comedies contain several such prayers, for example, (*sc. Iuppiter*) *propitius sit* (*Amph.* 935).[37] In the conclusion to book one of the *de Natura Deorum*, Cicero rejects the use of a similar prayer to the Epicurean gods: *deinde si maxime talis est deus ut nulla gratia, nulla hominum caritate teneatur, valeat—quid enim dicam "propitius sit?" esse enim propitius potest nemini* (1.124). Cicero mentions this particular prayer because it is an appropriate conclusion of an address to a god and thus contrasts well with *valeat*, which concludes a human conversation.[38] Petronius illustrates a different occasion for the use of a similar prayer. During Trimalchio's banquet, a slave enters the dining room carrying an offering of wine and shouting "*dii propitii*" (*Satyr.* 60). Servius *auctus* mentions such a practice as part of the regular domestic cult. Between the first two courses of a Roman dinner, there was a period of silence until a libation was offered at the hearth and a boy had proclaimed *dii propitii*

---

[34] This request lends weight to the suggestion that Aeneas is addressing Jupiter and not Mercury. Cf. Servius *ad* 4.577-578.

[35] Cf. later prayers that request that a god be *placidus* (Caes. Bassus *Carm.* 2.8; Sil. *Pun.* 8.142).

[36] I have found one example of *iuvare* with an adjective: *serenus | orsa iuves* (Val. Flac. 1.20-21).

[37] See also *Asin.* 781, 783, *Poen.* 1134; cf. *Merc.* 678, *Poen.* 278.

[38] See Pease on this passage under *valeat* and *propitius sit*.

(*ad Aen.* 1.730).[39] An asseveration that bears some similarity to the simple prayer appears twice in Cicero: *ita mihi deos velim propitios* (*Div. in Caecil.* 41; cf. *sim. Verr.* 5.37; also Petr. *Satyr.* 74). A Pompeian graffito reads *habeas propiteos deos* (CIL 4.4496g).

Livy uses the colloquial prayer *dii propitii sint* once when Aebutius' aunt responds to the troubles arising from her nephew's initiation into the Bacchic cult (39.11.7=App. 70). In addition to this formulaic usage, *propitius* also appears in the prayer of Horatius Cocles, where it is a transferred epithet describing the Tiber: *Tiberine pater . . . te sancte precor, haec arma et hunc militem propitio flumine accipias* (2.10.11=App. 9; cf. *flumine sancto* in the similar prayer of Verg. *Aen.* 8.72=App. 136).

secundus
Aen. 3.529, 8.302, 10.255

The only occurrences of this adjective in prayer appear in the *Aeneid.* Twice in the phrase *pede secundo* the adjective expresses the desire that deities (Hercules and Cybele) be propitious: *tua dexter adi pede sacra secundo* and *adsis pede, diva, secundo* (8.302, 10.255=App. 97, 102). Servius comments *ad Aen.* 8.302, *pede secundo*: *omine prospero.* In a third Vergilian prayer, the adjective retains the literal and more common meaning, "following," which derives from *sequor* (Ernout-Meillet, s.v. *secundus*). Anchises prays:

> di maris et terrae tempestatumque potentes,
> ferte viam vento facilem et spirate secundi
> (*Aen.* 3.528-529=App. 87).

Of course, a following wind is also a favorable one. Although *secundus* is not elsewhere attested in prayers, its usage in these Vergilian prayers to describe the gods is not novel (e.g. Liv. 7.26.7; cf. Enn. *Ann.* 255 *rumore secundo*). In addition, the adjective appears occasionally with reference to divination, especially augury (e.g. Enn. *Ann.* 79; Cic. *Div.* 1.27; Liv. 6.12.9). Both of these usages make its function in Vergilian prayers a natural, if not technical, one.

---

[39] *Apud Romanos etiam cena edita sublatisque mensis primis silentium fieri solebat, quoad ea quae de cena libata fuerant, ad focum ferrentur et in ignem darentur ac puer 'deos propitios' nuntiasset, ut diis honor haberetur tacendo* (Serv. *auct. ad Aen.* 1.730). See Friedlander for literature on this practice (313).

volens propitiusque

Liv. 1.16.3, 7.26.4, 24.21.10, 24.38.8, 29.14.13

The formula *volens propitius esse / fieri* first appears in a Plautine parody of a prayer and libation offered to a door:

> PH. Agite bibite, festivae fores;
> potate, fite mihi volentes propitiae (*Curc.* 88-89).

Later examples show that the formula was used in both private and public cult.[40] It occurs in three prayers that Cato records for use on his farm (*Rust.* 134.2-3, 139, 141), as well as in prayers written for the *Ludi Saeculares* (Pighi 114.96, 99=CIL 6.32323.96, 99; Pighi 157.12=CIL 6.32329.12). Three *leges ararum* also contain this phrase (*Lex Salonitana* CIL 3.1933=ILS 4907; *Lex Narbonensis* CIL 12.4333=ILS 112; *Lex Tiburtina* II 4.1, n.73=Palmer (1974) 58). It appears in a passage of Festus referring to the dedication of a *templum*: *in quibus (sc. pontificis libris), [scribtum est: "templum]que sedemque tescumque, [sive deo sive deae] dedicaverit, ubi eos ac[cipiat volentes] propitiosque"* (488L.= 356M.=*Gloss. Lat.* 4.446). In actual prayers, the request that a god be propitious is grammatically separate from the specific petition, for example, *quaeso precorque uti . . . rem . . . p(ublicam) . . . salvam servetis uti sitis volentes propitiae* (*Act. Lud. Saec.* Pighi 114.96=CIL 6.32323.96; cf. Macrob. *Sat.* 3.9.8; Cato *Rust.* 139, 141). The placement of the formula within prayers also may have been consistent; at least in public cult, it is always the concluding petition, following any specific requests. Among actual prayers, only in Cato's lustral prayer does this formula occur first. In two other Catonian prayers this is the only request.

The formulaic pair *volens propitius* occurs in five Livian prayers.[41] Two of these prayers combine the adjectives with the verb *adesse*, thus approximating the technical use with *esse* and *fieri*, discussed above. In the other three passages, Livy combines these adjectives with more active verbs: *sospitare, praebere, inire*. For example, when Romulus disappears before the eyes of his soldiers, they pray: *uti volens propitius suam semper sospitet progeniem* (1.16.3=App. 2). Here and in the other two passages (all in *oratio obliqua*), Livy may be conflating two clauses

---

[40] On *volens propitius*, see Servius *ad Aen.* 1.733 and Servius *auctus ad Aen.* 3.457, Norden (1939) 19 n.1, 285 n.91 and Robert E. Palmer (1974) 64.

[41] Cf. in non-prayer contexts, but of gods: Liv. 22.37.13, 39.16.11. The formulaic adjectives there contribute a solemn tone to official speeches.

in the interest of stylistic brevity; he combines the petition for divine favor with a second more specific request.

Pity

miserere

Aen. 9.495, 12.777

The verb *misereri* does not appear in the prayers of any Republican or Augustan prose author. Its earliest appearance in a prayer occurs in Catullus: *o di, si vestrum est misereri* (76.17).[42] In the *Aeneid*, both Euryalus' mother and Turnus offer this simple prayer for pity before they make more specific requests. For example, after the death of Euryalus, his mother prays:

> aut tu, magne pater divum, miserere, tuoque
> invisum hoc detrude caput sub Tartara telo
> (9.495-496=App. 101).

In non-prayer contexts, the verb also occasionally refers to divine pity (e.g. Afran. *com.* 417; Liv. 22.55.5).

miserescite

Aen. 8.573, 10.676

This synonym of *misereri* occurs in only two prayers in Latin literature, both Vergilian. Evander prays for the gods to "have pity" on him and, therefore, to preserve the life of his son Pallas:

> at vos, o superi, et divum tu maxime rector
> Iuppiter, Arcadii, quaeso, miserescite regis
> (8.572-573=App. 99).

Turnus prays for the winds to stop the ship which has carried him away from battle: *vos o potius miserescite, venti* (10.676=App. 104). *Miserescere* has no other religious usages in Republican or Augustan literature.[43]

B. Human Concerns

While great importance attached to the invocation of gods and pleas for divine favor, the primary objective of Roman prayers was to secure a

---

[42] This prayer is Catullus' only use of the verb. Vergil and Ovid use the verb of both human and divine pity (prayer: Ov. *Met.* 9.780, 11.133).

[43] For a later usage in epic, see Stat. *Theb.* 11.480. Cordier labels this a poetic word "found only in epic" (156).

safe and prosperous situation for mortals. This is the goal towards which other elements of prayer aim. Naturally, the specific requests made in prayers vary widely according to the situation and petitioner. Certain general concerns, however, reappear with some frequency and therefore make a useful framework for the subsequent discussion: the propitious outcome of events, preservation of safety, aid, aversion of danger, and curses.

Propitious Outcome

The majority of formulae and their variants request the propitious outcome of human affairs. Such petitions most frequently occur prior to some important undertaking such as a military expedition or individual battle. Many of these petitions indicate quite clearly the role of the gods by means of a divine subject or addressee. Some petitions, however, were abbreviated through the course of time, and to some extent secularized, so that the divine role receives little attention. One of the most common prayers in inscriptions as well as in literature, for example, is so abbreviated that it makes no direct reference at all to the gods: *quod bonum faustum felixque sit*. The religious language still makes it clear that this is a prayer, i.e. a request of the gods. Although a verb or verbal phrase such as *quod bene vertat* frequently carries the essence of this petition, a series of quasi-synonomous adjectives like *quod bonum faustum felixque sit* is common. These adjectives or their synonyms appear in several non-formulaic prayers in Livy's history.

Adjectival Constructions

quod bonum faustum felixque sit

Liv. 1.17.10, 1.28.7, 3.34.2, 3.54.8, 8.25.10, 10.8.12, 24.16.9, 42.30.10

A number of sources attest the predominantly official character of the formula *quod bonum faustum felixque sit* and its variations.[44] In his *de Divinatione*, Cicero tells us that this formula prefaced all undertakings: *quae (sc. omina) maiores nostri quia valere censebant idcirco omnibus rebus agendis "quod bonum faustum felix fortunatumque esset" praefabantur* (*Div.* 1.102). According to Varro, the *censoriae tabulae*

---

[44] The *Acta Fratrum Arvalium* and a censorial formula quoted by Varro expand the formula with *fortunatum* and *salutare* (Henzen 154, 8=CIL 6.32367.8; *Ling.* 6.86). Suetonius attests a briefer formula: *quod bonum . . . faustumque sit . . .* (*Aug.* 58.2).

prescribed the same prefatory prayer for summoning the people to the *lustrum* after a *census*:

> ubi noctu in templum censor auspicaverit atque de caelo nuntium erit, praeconi sic imperato ut viros vocet: "Quod bonum fortunatum felix salutareque siet populo Romano Quiritibus reique publicae populi Romani Quiritium mihique collegaeque meo, fidei magistratuique nostro" . . . (*Ling*. 6.86).[45]

The formula continues to appear in the religious language of the Augustan and later imperial periods. It appears in the dedication of an *ara Augusta* at Rome in A.D. 1 and of an *ara numinis Augusti* at Narbo in A.D. 11 (CIL 6.30975=ILS 3090, CIL 12.4333=ILS 112). In the *Acta Fratrum Arvalium*, this prayer frequently introduces the announcement (*indictio*) of sacrifices for the Dea Dia (e.g. a. 38: CIL 6.32344.1= Henzen XLII; a. 87: 6.32367.1=Henzen CXVI; see Henzen 8). The same prayer also precedes the cooptation of a priest to the brotherhood in some *acta* (Henzen CLIX.16; cf. 153-154). The appearance of this formula in abbreviated form, *q.b.f.f.*, in imperial inscriptions is in itself proof of its frequency (e.g. CIL 4.1354, 12.333). Finally, passages in Plautus and Apuleius suggest that this formula saw colloquial as well as official use (Plaut. *Trin*. 41; Apul. *Met*. 2.6).[46] In all of these examples, the formula stands as a single discrete prayer, prefacing the statement of an action.

The prefatory formula *quod bonum faustum felixque sit* appears eight times in Livy, without variation in wording except for the omission of the enclitic *-que* in one instance. All examples of the formula appear in an official setting, four of the eight in the context of a *contio* (1.17.10, 3.34.2, 3.54.8, 10.8.12=App. 4, 13, 15, 37). It is always spoken by people in official positions, a king, envoys, and senators (e.g. 1.28.7, 3.54.8, 42.30.10=App. 7, 15, 77), and to people carrying out official functions, such as soldiers and assemblies (e.g. 24.16.9, 3.54.8=App. 45, 18). Its objective is always a divine blessing on acts about to be undertaken that

---

[45] On Varro's sources, see Norden (1939) 4-5. Cf. Val. Max. 4.1.10 on the changes made by Scipio Africanus to the censorial carmen. Cf. *quod mihi reique fidei regno vobisque, Quirites, | se fortunatim feliciter ac bene vortat* (Enn. *Ann*. 102-103 Skutsch=107-108 V^3).

[46] Cf. Apul. *Met*. 11.20. See also the Pompeian graffito CIL 4. 1679.

will affect the state such as voting (e.g. 1.17.10=App. 4). Thus, although Livy does not present any of the occasions which occur in other sources, the historian clearly uses this formula in the same manner.

Synonyms

faustus

Liv. 2.49.7, 27.45.8

Outside of Livy, *faustus* only occurs in prayers in the formula *quod bonum faustum felixque sit* and its variants (q.v.).[47] In Livy, however, *faustus* appears in two non-formulaic prayers. In both, it occurs with its alliterative synonym *felix*, which accompanies *faustus* in the technical formula. As the Roman army marches past the Capitolium on its way to besiege Veii, well-wishers pray, *ut illud agmen faustum atque felix mittant* (2.49.7=App. 10). Again, before C. Nero's encounter with Hasdrubal, the Romans ask the gods, *ut illis faustum iter, felix pugna, matura ex hostibus victoria esset* (27.45.8=App. 50). There is no good evidence to suggest that either of these passages recalls a formulaic prayer.[48] In both situations, the author recalls the informal prayers of the masses; an official formula would be inappropriate here.

felix

Liv. 2.49.7, 21.50.8, 22.30.4, 27.45.8

On four occasions, Livy employs the adjective *felix* in phrases other than the formula *quod bonum faustum felixque sit*, none of which are attested elsewhere in prayers.[49] Once, *felix* appears alone in an apparent abbreviation of the official formula. After being saved from capture by the Carthaginians, M. Minucius Rufus addresses Q. Fabius Maximus: *quod tibi mihique servato ac conservatori quod exercitibusque his tuis sit felix* (22.30.4). In the characteristic fashion of the formula, this brief prayer introduces a statement: *sub imperium auspiciumque tuum redeo*. It also includes the formulaic *quod* and *sit*. Finally, the prayer employs the same datives of reference found in all but one occurrence of the technical

---

[47] *Faustus* does not occur at all in Vergil, but *infaustus* appears four times.

[48] But cf. Lucretius' reference to the sacrifice of Iphigenia: *exitus ut classi felix faustusque daretur* (Lucret. 1.100). See TLL s.v. *felix* 6.450.22-26. "*quasi in formula*."

[49] See Zieske 220-232 and Erkell 54-71 for discussion of the *felix* word family in Livy.

formulae in Livy.[50] Without exactly reproducing the official formula, the words lend a solemn and magisterial tone to Minucius' expression of gratitude. In two other prayers, *felix* appears in proximity to *faustus*, another adjective from the formula *q.b.f.f.* (2.49.7, 27.45.8=App. 10, 50; see above s.v. *faustus*). On one occasion, it occurs with its synonym *prosper*; Hiero requests a *prosper ac felix transitus* to Sicily for Ti. Sempronius (21.50.8=App. 43).

laetus
Liv. 37.47.5

The petition *ut . . . prospera et laeta eveniret* (Liv. 37.47.5=App. 66) is a unique variation on the formulaic phrase Livy regularly uses of propitiatory supplications (*bene ac feliciter eveniat* q.v.). Like its formulaic model, the variant also refers to the decree of a *supplicatio*; this thanksgiving took place after the encampment of Scipio Africanus' army in Asia. Elsewhere, Livy's own usage of the phrase *prospera et laeta* and its usage in Festus suggest that it derives from augural language. In reference to divination, Scipio tells an assembly of his soldiers that the gods portend success for their endeavors: *dii . . . auguriis auspiciisque et per nocturnos etiam visus omnia laeta ac prospera portendunt* (26.41.18). A similar phrase is used of the gods' favor toward a war with the Macedonianians; P. Sulpicius Galba tells an assembly of the people that, in response to his sacrifices and prayers (*ut hoc bellum . . . bene ac feliciter eveniret*), the gods *laeta omnia prosperaque portendere* (31.7.15=App. 58). The adjectival phrase also appears in fragments of the Augustan jurist Ateius Capito, as quoted by Festus: *stellam significare . . . laetum et prosperum; sinistrum in auspicando significare . . . laetum et prosperum auspicium* (Fest. 476L.=351M.).[51]

prosperus
Liv. 21.21.9, 21.50.8, 37.47.5, 42.28.7

Outside of Livy there is only one prayer that uses the adjective *prosperus*. In Gnaeus Gellius' account of the war between the Sabines and Romans, Hersilia prays that Neria Martis intervene: *te obsecro, pacem da, te uti liceat nuptiis propriis et prosperis uti* (*apud* Gell. *NA* 13.23.13).

---

[50] On the datives, see Appel 160 and Versnel (1976) 386 n.57. Cf. Liv. 41.16.2.

[51] *Laetus* also appears alone in augural contexts. On *laetus* in the language of augury, see L.R. Palmer (1954) 69-70. For additional examples, see TLL 7.888.63-889.14.

In Livy the adjective *prosperus* appears twice with the verb *evenire* in variations on his standard formula for propitiatory *supplicationes* before military expeditions (37.47.5=App. 66 with *laetus* q.v.; 42.28.7=App. 76). It appears again with *evenire* in a similar prayer, a vow Hannibal makes for divine favor on his military endeavors (21.21.9=App.120).[52] *Prosperus* also occurs with its synonym *felix* in Hiero's prayer for the journey of Sempronius: *precatusque prosperum ac felicem . . . transitum* (21.50.8=App. 43). The historian's usage may have been influenced by the technical use of the adverb *prospere* in reference to gratulatory *supplicationes* (e.g. Aug. *Anc.* 4.2; CIL 14.3613=ILS 918). Another influence on Livy's diction may have been the use of *prosperus*, like *laetus* (with which it is sometimes paired), as an augural term.[53]

Verbal Constructions

adesse

Liv. 2.6.8, 3.25.8, 6.29.2, 7.26.4, 24.38.8

Aen. 1.734, 4.578, 8.78, 10.255, 10.460, 10.774

The simple petition *ades*, with its sundry forms, varies in meaning from the literal request for divine presence to a request for divine aid. The two are not mutually exclusive; in all likelihood, the original concept was that physical presence was a necessary pre-condition for a divinity to give aid. In the only attestation from an actual prayer, one of the few prose examples, the usage of *adesse* is unclear since the petition occurs out of context. Servius comments on the Vergilian phrase *tuo genitor cum flumine*: *sic enim invocatur in precibus "adesto, Tiberine, cum tuis undis"* (*ad Aen.* 8.72).[54] There is insufficient information here to know whether the speaker invokes Tiberinus simply to attend to a prayer or ceremony or whether aid is being requested. The only other pre-Augustan example occurs in the repeated refrain from Catullus' *epithalamium*: *Hymen o Hymenaee, Hymen ades o Hymenaee* (62.5).

---

[52] S.v. *bene ac feliciter eveniat* for these three petitions. Cf. the use of the adverb in 7.11.4=App. 114.

[53] See above s.v. *laetus* for the combination *prosper et laetus*. For *prosper* alone, see Naev. *BPunic.* 34.2; Val. Max. 1.5.1; Plin. *HN* 10.21; Fest. 65L.=74M. For use of the adverb, see Gell. *NA* 7.6.8.

[54] Ogilvie (1965) refers to *adesse* as a "very ancient form of invoking the help of gods . . . or men" (375). Robert E. Palmer (1970) suggests that this invocation was part of an augural *precatio* (90). On this suggestion, see Linderski (1986a) 2253 n.418. For further discussion of *adesse*, see Bömer 14 with literature.

Here *ades* functions as a traditional hymnic invocation for the presence of the marriage deity at the nuptials. As part of an invocation, this verb sometimes requests a deity's presence in order to hear a petition, as clearly seen in this example from Ovid's *Metamorphoses*: *huc ades atque audi placidus, Neptune, precantem* (8.598).[55] Examples of requests for aid are common; one of the most frequent occasions is the poet-narrator's petition for divine inspiration (e.g. Verg. *Geor.* 1.18; cf. also Ov. *Fast.* 3.2; Quint. *Inst.* 4. *praef.*4).

In each of five prayers in Livy that contains *adesse*, the content and situation show that the speaker seeks divine aid and not simply divine presence.[56] All precede armed encounters, ranging from the individual attack of Arruns Tarquinius on Brutus (*di regum ultores adeste*: 2.6.8=App. 8) to a battle between the Roman and Praenestean armies (*adeste, di testes foederis*: 6.29.2=App. 27). None of these prayers includes an additional petition in verbal form. Twice, however, a request for divine favor is included through the predicated formula *volens propitius* (7.26.4, 24.38.8=App. 29, 47).

The *Aeneid* presents more varied examples of *adesse* in prayer than does Livy's history. Some seem to be mere requests for the presence of some quality associated with the god. At the banquet given for the Trojans, Dido expresses her wish for a joyous evening by invoking Bacchus, scarcely more than a metonomy for merriment: *adsit laetitiae Bacchus dator* (1.734=App. 79). A similar impression attaches to Aeneas' petition to the deity who sent Mercury to order his departure from Carthage:

> adsis o placidusque iuves et sidera caelo
> dextra feras (4.578-579=App. 89).

The request for favorable skies along with the predicate adjective *placidus* ("tranquil") suggests that Aeneas is invoking good weather rather than a deity. When Aeneas prays to the god of the Tiber *adsis o tantum et propius tua numina firmes* he is not so much interested in an

---

[55] Note W. S. Anderson's comments on *huc ades atque audi placidus, Neptune, precantem* (Ov. *Met.* 8.598). "Heinsius wished to reject this line. Although it exhibits a typical prayer formula, appropriate to the beginning of an invocation and so accepted by others, the writer probably did not fully visualize the scene. After all, Neptune *is* present as part of the water." But compare Vergil's use of *adesse* in prayers to gods who have just manifested themselves.

[56] Cf. the adjective *praesens* which often describes a god who is perceived as a helper (e.g. Liv. 29.18.12; Verg. *Aen.* 9.404=App. 100). See OLD, s.v., n.3.

epiphany (the god has just appeared to him in a dream) as a quick confirmation of the god's prophecy (8.78=App. 136). Following another divine vision, Aeneas prays for Cybele to bring close (*propinques*) her augury too and to assist the Trojans in their impending battle with the Rutulians: *adsis pede, diva, secundo* (10.255=App. 102). Mezentius certainly does not invoke the presence of his own right hand and weapons, but rather their assistance in combat:

> dextra mihi deus et telum, quod missile libro,
> nunc adsint! (10.773-776=App. 139; cf. Stat. *Theb.* 9.549).

augere
Liv. 29.27.4

The verb *augere* occurs only rarely in prayers. Twice Plautus combines the verbs *augere* and *adiuvare* in what appears to be a humorous conflation of three colloquial prayers (*di adiuvant, augent, amant*) where the petition seeks general divine favor (*Epid.* 192, *Men.* 551). In the official prayers of the *Ludi Saeculares* the basic meaning, "to increase," is visible: *uti imperium maiestatemque p. R. Quiritium duelli domique auxis/itis* (Pighi 114.94=CIL 6.32323.94, Pighi 157.11=CIL 6.32329.45).[57]

The phrase *bonis auctibus auxitis* (*auxeritis*) does not appear outside of Livy (29.27.4=App. 55; auctibus *Frob2*; auctoribus *cod.*). The prayer, which Scipio offers before setting out on his African campaign juxtaposes three quasi-synonymous verbs: *quae in meo imperio gesta sunt geruntur postque gerentur, ea . . . bene verruncent, eaque vos omnia bene iuvetis, bonis auctibus auxitis.* Although syntactically varied and etymologically distinct, all three petitions seek success as the result of divine favor. The derived noun *auctus*, which is here best understood as "success," affects the basic meaning of *augere*.[58] The phrase as a whole may be translated by "may you cause to prosper." Although similar phrases in Livy and other authors refer to literal increases, the object *ea*

---

[57] On the authenticity of this line, see Harris 121 n.6, 266; Momigliano 165; Diehl 357-369. For a discussion of prayers for the expansion of Roman territory, see Harris 118-123.

[58] Most modern scholars, as well as some ancients, postulate an etymological relationship between *augere* and *augur* and *augustus*. On etymology, see Dumezil 80-102, Wagenwoort (1947) 12-17, Erkell 22-24, Ernout 67-71, and Ernout-Meillet^4, s.v. Note also Ov. *Fast.* 1.609-612.

(past, present, and future acts of Scipio's command) demands a more abstract meaning here.[59]

bene ac feliciter eveniat
Liv. 21.17.4, 31.5.4, 7.15, 8.2, 36.2.2, 40.46.9
variants: Liv. 7.11.4, 21.21.9, 37.47.5, 42.28.7

The common propitiatory prayer in Livy, *bene ac feliciter eveniat*, and its variants appear in a number of prose authors as early as Cato. In his prayer for the lustration of a field, that author employs a briefer phrase using a single adverb: *utique tu fruges, frumenta, vineta virgultaque grandire beneque evenire siris* (*Rust.* 141.3). A similar prayer occurs in a pseudo-Sallustian letter: *di immortales adprobent beneque evenire sinant* (1.8.10). The single adverb *bene* also occurs in an abbreviated version of this prayer, (*quod*) *bene eveniat*, which came into colloquial use (Cato *Rust.* 141; Pomp. *com.* 35; Cic. *Att.* 7.2.4; Petr. *Satyr.* 99; CIL 4.4477). Cicero identifies the lengthier formula *bene ac feliciter eveniat* as official: the *sollemnis comitiorum precatio*. He recalls the formulaic prayer he offered on the day of Murena's election to the consulate: *quae precatus a dis immortalibus sum, . . . more institutoque maiorum . . . ut ea res mihi fidei magistratuique meo, populo plebique Romanae bene atque feliciter eveniret, . . .* (*Mur.* 1). In the next sentence he paraphrases: *ut eis . . . ea res fauste feliciter prospereque eveniret* (1). To judge from Pliny, this same formula continued to refer to consular appointment during the following centuries: *iam quod precatus es, ut illa ipsa ordinatio comitiorum bene ac feliciter eveniret nobis rei publicae tibi, . . .* (*Pan.* 72.1). The formula appears in non-official contexts as well. In one of his letters, Cicero uses the formula colloquially, in reference to his marriage to Publilia (*Fam.* 4.14.1; cf. 12.19.1). Petronius' characters, Encolpius and his companions, offer the same prayer before setting out on a part of their journey (*Satyr.* 117).

In Livy, the formula *bene ac feliciter eveniat* occurs six times, always in an official context. Five of these passages are senatorial decrees directing the Roman people to offer propitiatory prayers for wars about to be undertaken. A pattern emerges in which the Senate first directs the consuls on their inaugural day to offer prayers that the war under consideration *bene ac feliciter eveniat* (31.5.4, 31.7.15=App. 57, 58). After the

---

[59] Cf. *maximis auctibus crescere* (Liv. 4.2.2), *magnis auctibus auxissent* (Ps. Sall. *Epist.* 1.7.4), and possibly *magnis auctibus auxerunt* (Fronto 226).

Senate has voted for the war and the people have approved it, another decree using the same words directs the proclamation of a propitiatory *supplicatio* (21.17.4, 31.8.2, 36.2.2=App. 42, 59, 131).[60] These words belong not only to senatorial decrees, but also to the actual wording of prayers. Thus Livy writes that in 201 the Senate directed the consuls to offer a sacrifice to the gods *cum precatione ea, "Quod senatus populusque Romanus de re publica deque ineundo novo bello in animo haberet, ea res uti populo Romano sociisque ac nomini Latino bene ac feliciter eveniret"* (31.5.4=App. 57). In one passage, Livy shows his familiarity with the more general use of this formula in varied official situations. Q. Caecilius Metellus asks the newly elected censors of 179, who are personal enemies, to earnestly desire the request each of them will make *in omnibus fere precationibus verbis ut ea res mihi collegaeque meo bene et feliciter eveniat* (40.46.9=App. 73).

In addition to the standard formula employing the adverbs *bene ac feliciter*, Livy also creates unique variants by substituting other modifiers. Twice when referring to propitiatory prayers, Livy substitutes predicative adjectives for the adverbs; both passages include the adjective *prosperus* (q.v.). That adjective appears alone in the Senatorial decree that the consuls offer prayers concerning the anticipated third Macedonian war: *quod bellum populus Romanus in animo haberet gerere id prosperum eveniret* (42.28.7=App. 76). The phrase is expanded with *laetus* (q.v.) in a similar decree during the Syrian War: *ut ea res prospera et laeta eveniret* (37.47.5=App. 66).

Livy includes similar petitions in vows for military success. The adjective *prosperus* appears again with *evenire* in a vow Hannibal makes for divine favor toward his military endeavors: *se obligat votis, si cetera prospera evenissent* (21.21.9=App. 120; cf. adverb in 7.11.4=App. 114). The adverb appears with *evenire* in another vow made by Servilius Ahala: *si prospere . . . evenisset* (7.11.4=App. 114).[61] In both the supplications and vows, the historian's choice of *prosperus* and *prospere* may have been influenced by the contemporary technical use of the

---

[60] On procedures for declaring war, see Rich 6. On propitiatory supplications, see Halkin 10-12. Cf. the parallel expression for supplications of thanksgiving on the successful completion of a military endeavor *bene ac feliciter gerere* (Liv. 31.48.12; 39.4.2).

[61] In addition, *prospere* is combined with *bellare* and *gerere* in two other vows (22.9.10; 32.6.7=App. 122, 128).

adverb *prospere* in gratulatory supplications.[62] There is no evidence that its appearance in a few other situations is technical (e.g. Cic. *Mur.* 1; Ps. Sall. *Epist.* 2.13.8).

bene verruncent
Liv. 29.27.3

Verruncare seems to be a poetic synonym for *evenire* and *vertere.*[63] Unlike the prayer *bene ac feliciter eveniat* above, there is no evidence that *bene verruncent* is a common formula in either colloquial or official language. Before Livy, this rare verb appears only in prayers written by early playwrights. Two prayers from *praetextatae* by Accius both request that portents turn out propitiously for the Roman people: *portenta ut populo patriae verruncent bene* (5-6; cf. 36). A prayer from Pacuvius' *Periboea* contains the following fragment of a prayer, perhaps offered by Diomedes as he plans revenge against the usurpers of Oeneus' throne: *ut quae egi ago axim verruncent bene* (*trag.* 297).

The similarity of the latter prayer to that of Scipio before embarking for Africa is striking: *precor quaesoque uti quae in meo imperio gesta sunt geruntur postque gerentur, ea mihi populo plebique Romanae . . . bene verruncent, eaque vos omnia bene iuvetis, bonis auctibus auxitis* (Liv. 29.27.3=App. 55). Livy's prayer uses both the unusual verbal triad and the archaic verb *verruncare* found in Pacuvius' prayer. This is the only prose attestation of *verruncare.*[64] Its apparent allusion to an archaic model contributes to the general antiquarian character of Scipio's prayer. Although the objective of the prayer *bene verruncent* does not differ from that of *bene ac feliciter eveniat*, it serves a different function. Livy uses the latter formula only as a single brief prayer in a public setting. In contrast, the context for *bene verruncent* is a lengthy prayer spoken in front of soldiers about to embark on a long and dangerous expedition. This

---

[62] But cf. *ut ea res ei imperatori prospere feliciter cedat* (*Act. Frat. Arv.* a. 213, Henzen CXCVII.22=CIL 6.2086.22). Livy follows the technical usage of the Augustan period by preferring *prospere gerere* in reference to *supplicationes* of thanksgiving (26.21.3, 33.24.4, 35.5.5, 35.36.12, 39.38.5, 40.16.4, 41.12.5, 41.17.3, 41.28.1). *Prospere* is one of three adverbs (with *bene* and *feliciter*) which appear in decrees and prayers of *supplicationes* giving thanks for military victories. During the Augustan period, it is the preferred term (Aug. *Anc.* 4.2; CIL 14.3613=ILS 918; *et al.*).

[63] Paulus Festus glosses *verruncent* as *vertant* (511L.=373M.).

[64] The vow offered at Trajan's *profectio* in 101 does show a similar verbal pattern, but refers only to present and future: *earumq(ue) rerum ei, quas nunc agit actur[usve est, bonum eventum] des* (Henzen CXL.29=CIL 6.2074.29).

prayer carries the burden not only of seeking divine favor, but also of encouraging and inspiring its human audience. Therefore, a more rhetorical language seems appropriate.

bene vertat
Liv. 1.28.1, 3.26.9, 35.8, 62.5, 7.39.13, 8.5.6, 34.34.2
di bene vertant 10.35.14, 29.22.5; *Hercules* 10.18.14

The only occurrence of the phrase *bene vertat/vortat* in an official context prior to Livy is found in Ennius' *Annales*:[65]

> quod mihi reique fidei regno vobisque, Quirites,
> se fortunatim feliciter ac bene vortat
> (102-103 Skutsch=107-108 V^3).[66]

From Nonius' information that the fragment derives from book one and from the use of the word *regno*, it is reasonable to assign these words to the king Romulus. The address of the Roman people as *Quirites* further confirms that the occasion is an official one. According to Livy and other sources, the Romans adopted this name, derived from the Sabine town of Cures, after the union of the two peoples.[67] Nonius preserved these two lines without further context, and it is not possible to say definitively in what context these words appear. If, however, Livy's usage is a valid guide, Romulus speaks them prior to making an official announcement to the people, as a *praefatio*. Based on the first occurrence of the formula in Livy, one could also speculate that the nature of that announcement was the union of the Romans and Sabines. The victorious Roman king Tullus Hostilius prefaces his order for the union of the Alban and Roman camps with the same formula: *quod bene vertat, castra Albanos Romanis castris iungere iubet* (1.28.1=App. 6).

Besides this one Ennian example, similar prayers in authors before Livy occur only in the comic poets and are restricted to private individuals and situations. On a few occasions, however, prayers may mimic official practice, or alternatively, reflect a colloquial usage similar to the official. There are three petitions in Plautus that clearly function as

[65] Cf. curses with *vertere* (e.g. Plaut. *Curc.* 273; Ter. *Phorm.* 678). On expansion of this *precatio detrita*, see Appel 172 and Weissenborn-Müller's edition of Livy (1908) on 1.28.1.

[66] On this and similar *praefationes* and prayers, see Skutsch (1985) 250-252.

[67] Liv. 1.13.5. On the term *Quirites*, see the recent article by Prugni with ample literature (127-161).

praefationes to other statements. No doubt it is with a touch of parody that Plautus commences the preface itself of the *Asinaria* with such a *praefatio*. When the poet asks the audience to listen to the words of the herald, he is probably mimicking the official practice of asking for a propitious outcome of the words to be spoken:

hoc agite sultis, spectatores, nunciam,
quae quidem mihi atque vobis res vortat bene
gregique huic et dominis atque conductoribus.
face nunciam tu, praeco, omnem auritum poplum (1-4).

A similar *praefatio* occurs before the issuing of directions to a slave in the *Captivi* (361). Finally, Plautus clearly parodies Roman prayer in the words spoken by Saturio before the parasite exhorts his daughter to cooperate in his scam to sell her:

SAT. quae res bene vortat mihi et tibi et ventri meo
perennitatique adeo huic, perpetuo cibus
ut mihi supersit, suppetat, superstitet: . . . (*Pers*. 329-331).

In addition to these clear examples of *praefationes*, Plautus' comedies contain another more numerous group of petitions, which originally may have functioned in the same manner. Plautine usage suggests that the petition *bene vortat* was the customary prayer made on the occasion of a betrothal; it accompanies all but one such event in that author.[68] Unlike the three prayers previously discussed, these may occur either before or after the exchange of promises. The following example of a betrothal also shows a variant of the petition in which the gods are the subject of the sentence and the verb takes on the meaning "to cause to turn out":

PH: . . . nunc tuam sororem filio posco meo.
quae res bene vortat! quid nunc? etiam consulis?
LE: quid istic? quando ita vis: di bene vortant! spondeo
(*Trin*. 571-573).

Other propitiatory prayers in the same context are sometimes expanded with datives, thus creating a more formal tone, for example:

LY. . . . nunc quae res tibi et gnatae tuae
bene feliciterque vortat—ita di faxint, inquito
(*Aul*. 787-788).

---

[68] But Müller emends *evenisse* in *Poen*. 1078 to *vertisse* (528-529). On this betrothal prayer, see also Köhm 34-35.

In contrast to the colloquial use of the formula in the comic authors, Livy, perhaps following Ennius' lead, consistently restricts this petition to official situations.[69] The speaker always seeks divine favor on an action about to be undertaken by public figures. With one exception, the persons making the petition act in an official capacity such as king, envoy, or consul (e.g. 1.28.1, 8.5.6, 10.18.14=App. 6, 32, 39). The same prayer offered by generals announcing military decisions is even used by a mutinous cohort that decides to summon Cincinnatus as its leader (7.39.13=App. 30; cf. 3.62.5=App. 16). The human audience also has some public role, for example, senators, a general, or soldiers (e.g. 8.5.6, 29.22.5, 3.62.5=App. 32, 54, 16).

Livy's usage of the formula *bene vertat* is very similar to that of *quod bonum faustum felixque sit.* Rather than complete and distinct prayers, both function as brief almost parenthetical remarks preceding statements.[70] Despite the colloquial contexts that comedy evidences for the former of these petitions, Livy does not always make an obvious distinction between the two formulae on the basis of official versus colloquial contexts. For example, both formulae preface addresses to soldiers (e.g. 10.35.14, 24.16.9=App. 41, 45). The most consistent correlation between the formula used and the context is *quod bonum faustum felixque sit* and the identification of an assemblage as a *contio.* Of eight instances when Livy uses that formula, he identifies six of the audiences with the term *contio.* On two other occasions the speaker uses the formula when addressing consuls. None of the audiences before whom speakers pray *quod bene vertat* are identified as *contiones.*

Another interesting observation on Livy's usage of these two formulae is a consistently alternating pattern in those books that contain more than one example of either formula, with the exception of book ten. All the books that contain more than one of these formulae fall into the first decade. This pattern of *variatio* is illustrated by the following table:

---

[69] Cf. Nero's words at the beginning of work on the Corinthian canal: *sibi ac populo R(omano) bene res verteret* (Suet. *Nero* 37.3).

[70] Ogilvie (1965) refers to the formula *quod bene vertat* as a "pious aside" (442).

Table I: Variation of *Praefationes*[71]

Reference	***quod bonum faustum felixque sit***	***quod bene vertat***
1.17.10	*	
1.28.1		*
1.28.7	*	
3.26.9		*
3.34.2	*	
3.35.8		*
3.54.18	*	
3.62.5		*
8.5.6		*
8.25.10	*	

By alternating usage of the two formulae, Livy retains an official tone without undue repetition. This motivation also partially accounts for his use of a more colloquial and archaic formula alongside an official and contemporary one. The desire to recall the work of Ennius may have been another significant factor.

Several variations on the basic formula occur in Livy, as in his predecessors. In three of the ten passages that contain this petition, Livy uses the causative construction (10.18.14, 10.35.14, 29.22.5=App. 39, 41, 54).[72] In the other seven petitions, he employs the intransitive construction found in Ennius (only 34.34.2=App. 61 lacks *quod*). On only two occasions does Livy expand the petition with datives expressing the intended beneficiaries, as does the Ennian prayer, *Annales* 107 (3.26.9, 8.5.6=App. 12, 32).

[71] One passage in book ten does not follow the pattern: 10.8.12: *q.b.f.f.*, 10.18.14: *q.b.v.*, 10.35.14: *q.b.v.* In other books, there is no more than one occurrence of either petition.

[72] Hercules is specifically invoked in 10.18.14.

iuvare
Liv. 29.27.3
Aen. 4.578

The technical verb *iuvare* appears in the oldest surviving Latin prayer, the archaic *carmen Arvale*: *enos Lases iuvate* and *enos Marmor iuvato* (Henzen CCIV.32-37=CIL 6.2104.32-37). The verbal phrase *bene iuvare* occurs in a number of other prayers in a variety of circumstances. In a third century inscription found at Falerii, it is part of a guild dedication: *ququei huc dederu[nt i]nperatoribus summeis, utei sesed lubentes bene iovent optantis* (CIL 1^2 364=ILS 3083=CE 2).[73] Plautus uses the phrase in a lengthy parody of the prayer of thanksgiving spoken by a triumphing general: . . . *quom bene nos, Iuppiter, iuvisti* . . . (*Pers.* 753-755). Since there was a close relationship between propitiatory prayers offered prior to a military expedition and gratulatory prayers offered upon return, the Plautine prayer of thanksgiving may point to the usage of *iuvare* in the earlier propitiatory prayers as well.[74] Describing the *precatio maxima*, which was recited at the *augurium salutis*, Servius *auctus* also employs the verb *iuvare*: *ut melius iuvent, melius fortunent* (*ad Aen.* 12.176).[75] Finally, it is worth noting that the phrase was probably used in augural language to indicate an auspicious bird: *bene iuvante ave.*[76] In addition to these formal usages, *iuvare* appears in the colloquial oath *ita me di iuvent*, which functions as an asseverative (Cic. *Att.* 1.16.1; Catull. 61.196, 66.18).[77]

This collection of precedents shows that Livy uses traditional language in the formal prayer of Scipio Africanus: *uti quae in meo imperio gesta sunt geruntur postque gerentur, . . . eaque vos omnia bene iuvetis* (29.27.2-4=App. 55). The precedents also show, however, that the historian deviates somewhat from the syntactic pattern. Although all of the extant examples from predecessors show that a person is the object of the request, Livy's prayer makes human actions the object.

---

[73] On this inscription see Peruzzi 115-162; Pisani 17-18; Linderski (1958) 47-50.

[74] For a discussion of the Plautine prayer, see Fraenkel (1922) 236-240.

[75] On the *precatio maxima* and the *augurium salutis*, see Linderski (1986a) 2254.

[76] See Linderski (1986a) 2285. Cf. *dis bene iuvantibus* (e.g. Liv. 6.23.10; cf. Cic. *Phil.* 3.36 and "with God's help").

[77] For additional examples of *iuvare* and *adiuvare* in prayers, see Appel 127.

Livy's usage may contribute to a better understanding of the meaning of the verb in those ealier passages. Because Scipio seeks a divine blessing on the actions of his command, past, as well as present and future, it is clear that the commander is not simply asking that the gods "help," as would be suggested by the basic meaning of the verb *iuvare*. Thus here as often in the earlier prayers, the translation "to cause to prosper" is desirable. The other requests in Scipio's prayer support this interpretation: *bonis auctibus auxitis* and *bene verruncent* (qq.v.). So, too, does the wording of the *precatio maxima*, which juxtaposes the requests *iuvent* and *fortunent* (see above).

Vergil also uses the verb *iuvare* in one petition, but without the formulaic adverb *bene*; instead the adjective *placidus* modifies the divine subject: *adsis o placidusque iuves* (4.578=App. 89).[78]

prosperare
Liv. 8.9.7

The first attested use of *prosperare* in prayer occurs in Livy, in the *carmen devotionis* of Decius Mus: *vos precor . . . uti populo Romano Quiritium vim victoriamque prosperetis* (8.9.7=App. 115). There is no solid evidence indicating that this is a technical term. No comparable phrase appears in the similar prayer found in Macrobius. Nor does the verb *prosperare* appear in the prayer that provides the closest linguistic parallel, that from the *Ludi Saeculares*: *victoriam valetudine[m populo Romano Quiritibus tribuatis faveatis]* (Pighi 114.95=CIL 6.32323.95). The only documentary source which attests its usage in prayer is a fragmentary metrical inscription of the Diocletianic period, addressed perhaps to the gods of Lavinium: *[p]rosperetis eventus* (CIL 14.2065, 2066=CE 212.6=ILS 6181, 6182). *Prosperare* does occur in a hymn written by Livy's contemporary Horace for performance at the *Ludi Saeculares* of 17 B.C.; the chorus prays Ilithyia to prosper the Augustan marriage laws: *diva . . . prosperes decreta* (1.17-18).[79]

---

[78] I find only one other example of *iuvare* with an adjective: *serenus . . . iuves* (Val. Flac. 1.20-21). Cf. *placidi servate* (*Aen.* 3.266=App. 86).

[79] Tacitus uses this verb twice to describe prayers (*Ann.* 3.56, *Hist.* 4.53). Cf. a poetic dedication of the late second century A.D. dedication: *Silvane . . . quod nos . . . tuo favore prosperanti sospitas* (CIL 12.103=CE 19.7).

secundare

Aen. 3.36, 7.259

Vergil is the first attested author to use the verb *secundare* ("to cause to be favorable") in prayers or even with deities as subjects. The latter usage first appears in the *Georgics*, where Cyrene tells her son, Aristaeus, that by binding Proteus he can force the god to explain the sickness of Aristaeus' bees and to bring about a favorable outcome of this event: *eventusque secundet* (4.397). Twice in the *Aeneid*, similar requests are made in the context of prayers. After seeing blood drip from tree roots he has just pulled up, Aeneas prays to the Nymphs and Mars that they bring about a favorable outcome of this portent: *rite secundarent visus omenque levarent* (3.36=App. 84). These words provided a model for two later prayers that visions be favorable (Luc. 1.635; Sil. 8.124). Another Vergilian prayer also refers to portents; King Latinus recalls signs and an oracle foretelling the marriage of his daughter to a stranger and the glorious destiny of their offspring:[80]

> . . . di nostra incepta secundent
> auguriumque suum! (*Aen.* 7.259-260=App. 96).

This association with omens recalls the technical usage of the related adjective *secundus* (q. v.) in augural language to describe *aves*.

Preservation

The verbs *servare, sistere*, and *sospitare* all appear in petitions seeking preservation. Although some of the petitions are made in times of danger, this is not always the case. Some petitions occur in ceremonies scheduled at regular intervals, with no regard for immediate need. Thus the urgent notion, which often attaches itself to the English word "to save," is not always appropriate. A more helpful translation, particularly when the adjective *salvus* accompanies the verb, is "to preserve."

servare

Liv. 22.10

Aen. 2.702, 3.86, 3.266

In actual Roman prayers, the technical term *servare* almost always appears in combination with the adjective *salvus*. In his prayer for the annual lustration of a private farm, Cato includes the petition: *pastores*

---

[80] Other examples of prayers using *secundare* are Stat. *Theb.* 1.59, *Achill.* 1.738, Sen. *Herc. Fur.* 645, Sil. *Pun.* 8.228.

pecuaque salva servassis (*Rust.* 141). The formula also appears in official prayers (e.g. Macrob. *Sat.* 3.9.11; *siritis*: *servetis emend. Büchler*). In prayers offered at the *Ludi Saeculares*, the same formula requests preservation of the state, for example, *remque publicam p. R. Q. salvam serves* (Pighi 116, 157=CIL 6.32323.129, 6.32329.12).[81] Vows made by the *Fratres Arvales* for the well-being of the emperor and his household, which probably recall Republican vows for the well-being of the state, also use the formula: *si . . . salvos servaveris* (Henzen 100, 103). In addition to this formula, the comic authors attest a colloquial expression that does not include the adjective *salvus*: *di servent* (e.g. Afran. *com.* 132; Plaut. *Poen.* 1258).[82] The acts of the Arval Brotherhood from A.D. 213 repeatedly record this same prayer for the emperor: *di te servent* (Henzen CXCVII=CIL 6.2074 a. 213: 17, 18, 19).

Livy uses the traditional adjective and verb combination in the proposal for a *ver sacrum*: *si res publica populi Romani Quiritium . . . salva servata erit hisce duellis* (22.10.2=App. 123). He has, however, used a passive contruction.[83] As in the examples of Livy's predecessors, the formula appears in an official context; the vow of a *ver sacrum* sought protection for the state after the horrible defeat at Trasimene.

None of the three Vergilian prayers containing *servare* uses the alliterative formula of the official prayers. But neither are the situations official ones. For example, Anchises anticipates the destruction of Troy when he prays: *di patrii; servate domum, servate nepotem* (2.702; cf. 3.86, 3.266=App. 83, 85, 86).

sistere
Liv. 29.27.3

Livy's history is the first extant text containing a prayer with the verb *sistere* and an adjective meaning "safe" as a variant of the formula *salvum servare*. Prior to his departure from Lilybaeum for the invasion of Africa, Scipio Africanus prays: *salvos incolumesque . . . mecum domos reduces sistatis* (29.27.3=App. 55). The only other prayer which contains

[81] On the phrase and its use in the *Ludi Saeculares* and Cato, see Diehl 365.

[82] For additional examples, see Appel 176-177. On *di servent* in Plautus, see Hanson 74-77. Cf. Jul. Max. 16.3.3 and *passim*, also Hor. *Carm.* 3.21.5.

[83] Cf. the explanation of a sacrifice of thanksgiving: [*quod dominus noster ex naufragii periculo s*]*alvus servatus sit* (*Act. Frat. Arv.*, Henzen CC.8=CIL 6.2103a 1.8).

a similarly worded petition is a public vow made on the occasion of Trajan's setting out to wage war against the Dacians and preserved in the *Acta Fratrum Arvalium*. More than a century after Livy (101), its witness is significant as a documentary example of the prayer offered at a *profectio*: [*eumque reduce*]*m incolumem victoremq*(*ue*) . . . *in urbem Romam sis*[*tas*] (Henzen CXL.31=CIL 6.2074.31). There is a contemporary example of the phrase *salvum sistere* in a vow, which Suetonius attributes to Augustus: *ita mihi salvam ac sospitem rem p*(*ublicam*). *sistere in sua sede liceat* (*Aug*. 28.2). This prayer is particularly interesting because Augustus prays that he himself may be the agent of what was traditionally seen as a divine gift. It is possible that Livy's usage reflects a contemporary development in religious language.

sospitare
Liv. 1.16.3

Prior to Livy, all examples of this verb are found in poetry and, with the exception of a Catullan poem, come from early Latin sources (Catull. 34.24). Fragments from Ennius and Pacuvius containing *sospitare* may be parts of prayers (Enn. *scen*. 295; Pacuv. *trag*. 234). A certain example of a prayer appears in Plautus: *di . . . istuc sospitent quod nunc habes* (*Aul*. 545-546).[84] In Livy's history the verb occurs in a prayer evoked by Romulus' apotheosis; the soldiers request his favor upon and protection of the Roman people: *volens propitius suam semper sospitet progeniem* (1.16.3=App. 2). This prayer resembles an Ennian fragment which also addresses Romulus (*Ann*. 110-114). Since Ennius is one of only three other authors in whom this verb is attested, it is reasonable to wonder whether a missing part of the Ennian prayer to Romulus contained the verb *sospitare*, and thus was a direct model for Livy's usage here.

Aid

There are two brief exclamations beginning with the word *pro* which trace their origin to prayers requesting the aid of the gods.

---

[84] Other examples are Plaut. *Asin*. 683, and Lucil. 739. On its usage in Livy, see Gries 61. Cf. also Juno's epithet *Sospes*.

pro deum fidem

Liv. 3.67.7, 44.38.10

Although the exact wording of the formula *pro deum fidem* does not occur in Plautus, some related variants that provide insight into its development and meaning do.[85] The fullest expression is *pro dei immortales opsecro vostram fidem*, in which *vostram* refers to the gods (*Poen.* 967; cf. *di, opsecro vostram fidem . . . servate nos*: *Cist.* 663-664). *Pro* is an interjection here accompanied by the vocative. It is clear that the speaker is invoking the gods and that the purpose is to request their aid. More often the Plautine formula is simply *di vostram fidem* (*Poen.* 830, 900, 953, cf. *Iuno Lucina, tuam fidem* (*Aul.* 692). That this is also a request addressed to the gods is well-illustrated by its appearance as the conclusion of a longer prayer spoken by Hanno:

> . . . deos deasque veneror qui hanc urbem colunt
> ut quod de mea re huc veni rite venerim,
> measque hic ut gnatas et mei fratris filium
> reperire me siritis, di vostram fidem! (*Poen.* 950-953).

These examples from Plautus show that the exclamation *pro deum fidem* was once a genuine appeal to the gods for help. Plautus also provides examples in which the formula is simply an asseverative exclamation, for example: *dei vostram fidem, nimium lepidum memoras facinus* (*Poen.* 900). Donatus comments on the same expression in Terence: '*di vestram fidem' admirantis adverbium est cum exclamatione . . . fidem dixit opem et auxilium* (*ad And.* 716). In Plautus and other authors, the formula almost always appears in the context of a question or an exclamation.[86]

The earliest appearance of the similar asseveration, *pro divum fidem* occurs in Ennius (*Sat.* 18). Terence also uses some variants on this formula: *di vostram fidem, pro deum immortalium,* and *pro deum atque hominum fidem* (e.g. *And.* 237, 246).[87] Cicero uses the expanded form, *pro deum hominumque* or *atque hominum fidem,* five times in his orations, twice in the dialogues, but never in the letters. Among the

---

[85] Secknus collects and discusses these *Ausrufe* (30). He identifies them as common Latin expressions without parallel in Menander.

[86] See Gagnér 198-204; TLL 6.666.29-64. s.v. *fides*.

[87] See Secknus 44-46 for a collection of these formulae. See also Caecil. *com.* 211.

historians it appears only once, in a speech attributed to Catiline by Sallust (*Cat.* 20.10).

Both occurrences of *pro deum fidem* in Livy appear in the context of questions within speeches (3.67.7, 44.38.10=App. 149, 166). Thus Livian usage is consistent with that of predecessors. For example, in a highly rhetorical speech regarding civil strife, which threatens the welfare of Rome, the consul T. Quinctius Capitolinus asks: *pro deum fidem, quid vobis voltis* (3.67.7=App. 149).[88]

pro Iuppiter
Aen. 4.590

Like the formula *pro deum fidem, pro Iuppiter* is composed of the interjection *pro* followed by a noun. Here the noun is the vocative form of Juppiter, the chief god of oaths. The exclamation is first attested in a fragment from Ennian drama (*scen.* 187). Plautus only uses this exact formula once (*Aul.* 241). On one other occasion, the playwright employs an expanded form with *supreme* (*Poen.* 1122). He prefers the expression *di immortales*.[89] Terence, however, prefers the invocation of Jupiter over that of the assembled gods. *Pro Iuppiter* appears five times and the synonymous *o Iuppiter* eight times (e.g. *And.* 464, 732). This is one of very few colloquial expressions appearing in the *Aeneid*. When Dido sees that Aeneas and his Trojans have set sail from Carthage, she begins the complaint, which ends in a curse of Aeneas and the future Romans, with the words *pro Iuppiter* (*Aen.* 4.590=App. 169).

Aversion of Danger

In addition to requests for propitious happenings, the Romans also petitioned the gods to ward off undesirable events such as defeat, famine, disease, and death.[90] Obviously such apotropaic prayers were common in times of trouble. They were also frequent in conversation when some undesirable event was mentioned. The mention of evil was in itself potentially effective in bringing it about. Likewise, the spoken apotropaic exclamation harnessed power to turn it aside. Here is clearly manifest one

---

[88] See Ogilvie (1965) 516-517 on Quinctius' speech and *ad loc.* on this formula, which he labels "an archaic exclamation."

[89] On the usage of Plautus and Terence, see Secknus 30 and 44-45.

[90] On apotropaic prayers, see Köves-Zulauf, who includes the petition *parce* with this type (69-70).

of the fundamental characteristics of Roman prayer, belief in the power of the spoken word.

arcere
Liv. 1.12.5, 5.18.12
Aen. 8.73

The verb *arcere* often conveys a specialized notion, in Ogilvie's words, "of keeping *profani* at a distance."[91] Thus at the close of his first oration against Catiline, Cicero turns to the statue of Jupiter Stator and addresses the god thus:

> tu, Iuppiter, qui isdem quibus haec urbs auspiciis a Romulo es constitutus, quem Statorem huius urbis atque imperi vere nominamus, hunc et huius socios a tuis ceterisque templis, a tectis urbis ac moenibus, a vita fortunisque civium omnium arcebis . . . (1.33).

By recalling the divine auspices with which the city was founded and mentioning the *templa* that stand within that city, Cicero points out that it is in the god's own interest to turn aside the profane conspirators. Cicero's wording bears a certain similarity to the prayer Livy places in the mouth of Romulus as he vows the temple of the same deity, Jupiter Stator: *Iuppiter, tuis . . . iussus avibus hic in Palatio prima urbi fundamenta ieci . . . at tu, . . . hinc saltem arce hostes* (1.12.5). The invading Sabines threaten to violate the holy place where Romulus had first taken the auspices (*tuis avibus*); they must be kept away (*arce*). In a second prayer, Livy's words recall the triple-noun phrase that describes the city of the Ciceronian oration. The Roman women, concerned about the safety of their city during the siege of Veii, pray to the gods *ut exitium ab urbis tectis templisque ac moenibus Romanis arcerent*. (5.18.12=App. 20). In the one Vergilian prayer using *arcere*, this idea of the separation of sacred and profane is not present. Before Aeneas seeks Evander's aid in war against the Latins, he prays to the nymphs and the Tiber river: *Aenean . . . arcete periclis* (*Aen.* 8.73=App. 136). Here there is only the notion of protection.

Despite its use in religious contexts, *arcere* does not belong among the technical terms of Roman prayer; actual prayers avoid this verb in

---

[91] Ogilvie (1965) 78 *ad l.*; to which add in non-prayer contexts e.g. Verg. *Aen.* 7.779; Pacuv. *trag.* 305; Tac. *Hist.* 5.8; Plin. *HN* 2.159. See also Appel 126. In Ennius, the earliest witness, the verb means *coerceo* (*Ann.* 543); see the TLL article 1.442.53.

favor of *avertere* and *prohibere*. In addition, usage patterns suggest a poetic tradition, at least during the Republic, which may account for its avoidance by Republican historiographers, including Cato, Caesar, Sallust, and Nepos. Although missing in Plautus and Terence, it does appear in Pacuvius and Accius (Acc. *trag.* 173; Pacuv. *trag.* 305).

avertere

Liv. 5.18.12, 23.13.4, 28.41.13
Aen. 3.265, 3.620

References in three authors attest the technical usage of *avertere* in formal prayers requesting that the gods "turn aside" some trouble.[92] A comment from Servius *auctus* suggests that this verb occurred in an augural prayer, perhaps spoken at the *augurium salutis*: *precatio, uti avertantur mala* (*ad Aen.* 3.265).[93] Often the trouble that gives rise to such a petition is a disease, as in the ancient prayer whose wording Arnobius has preserved: *saeva contagia et pestilentes morbos ab aestivis avertere* (*adv. Nat.* 3.23).[94] Another request for the aversion of disease appears in a prayer Festus gives for the lustration of a farm: *avertas morbum mortem labem nebulam impetiginem* (230L.=210M.). A reference in Livy to consultation of the Sibylline books also mentions the need to turn aside disease: *placandae deum irae avertendaeque a populo pestis causa* (4.25.3). In Romulus' prayer to the gods to defend Rome against the Veientines, *avertere* takes on an expanded meaning of "to turn away and transfer elsewhere." Thus Romulus prays: *ut exitium . . . arcerent Veiosque eum averterent terrorem* (Liv. 5.18.12=App. 20).[95]

Besides this formal usage, several passages from Ciceronian works, both orations and letters, attest the colloquial use of the apotropaic formula *di omen avertant* to ward off the adverse effects of naming potential

---

[92] Cf. the participle *avertentes* as a divine epithet (Macrob. *Sat.* 3.20.3; CIL 13.5197). Cf. also CIL 6.105, 106=ILS 3984.

[93] Possibly identical with the *precatio maxima*; see Linderski (1986a) 2253-2256.

[94] "Aus guter (wohl varronischer . . .) Uberlieferung," Norden (1939) 290, in reference to 249 n.3.

[95] See Ogilvie (1965) on this prayer. Cf. *pestem ab suis aversam in hostis* (Liv. 8.9.10). Cf. also Miss. Rom. Introit for the ninth Sunday after Pentecost: *averte mala inimicis.* For a brief discussion of *Gebetsegoismus*, see Versnel (1981a) 18-20 with references and Stübler 47.

difficulties (e.g. *Mur.* 88, *Phil.* 3.35, *Fam.* 12.6.2).[96] The same formula appears twice in Livy's history. Before mentioning possible adverse military situations, both Hanno and Q. Fabius beg the gods to turn aside the inauspicious omen of their words. For example, Q. Fabius, speaking to the Senate, prefaces the ill-omened statement *si . . . victor Hannibal ire ad urbem perget* with the formula *quod omnes di omen avertant* (28.41.13=App. 52).

Two Vergilian prayers bear certain resemblances to both the formal and colloquial petitions mentioned above. After hearing the prophecy of hunger by the harpy Celaeno, Anchises prays: *di prohibete minas; di, talem avertite casum* (3.265=App. 86). While the actual object of the verb *avertere* is *casum*, the timing immediately following the harpy's words shows that words of ill omen provide the motivation. A similar context surrounds the exclamation of Achaemenides, a lone Greek stranded on the island of the Cyclops, who prays *di talem terris avertite pestem*! when he mentions the monster (3.620=App. 88).[97]

di meliora

Liv. 9.9.7, 39.10.2

This colloquial exclamation ("may the gods grant more propitious events"), often expanded with a verb such as *faciant/faxint*, is common in both prose and poetry (e.g. Cic. *Cato M.* 47; Verg. *Geor.* 3.456, 513).[98] Twice in Livy's history, people express feelings of shock with this formula. In an emotional speech to the Senate, Sp. Postumius voices the horror inspired by an imaginary and exaggerated series of treaty conditions with the Samnites: *di meliora, inquis* (9.9.6=App. 35). Hispala also reacts thus to Aebutius' announcement that he will be initiated into the Bacchic rites (39.10.2=App. 68).

The comparative form of *bonus* appears in official as well as in colloquial prayers. Valerius Maximus writes of the archaic prayer offered at the close of a census for divine favor toward the Roman state, *ex publicis tabulis sollemne ei precationis carmen . . . quo di immortales ut*

---

[96] For additional examples in Cicero and later authors, see TLL 2.1323.3-6. Cf. the similar exclamation in Lucilius: *deum rex avertat verba obscena* (288).

[97] There are no other examples from the Augustan period. For later examples, see Appel 126, 170-171, and TLL 2.1322.83-1323.22.

[98] For additional examples, see Appel 174 and TLL 2.2092.78-80. For Greek parallels, cf. 'Ἀθῆνᾶ κρείττων' (Theophr. *Char.* 16) and Secknus 62. Cf. the adverbial parallels *di melius* (*faciant*) (TLL 2.2121.84-2122.7).

populi Romani res meliores amplioresque facerent rogabantur (4.1.10).[99] According to Servius *auctus*, *melius* also occurred in the *precatio maxima*, which was offered at the *augurium salutis*:[100] *eventusque rei bonae poscitur, ut in melius iuvent* and *quod in precatione dici solet, ut melius iuvent, melius fortunent* (*ad Aen.* 12.176). The prayers for the Augustan *Ludi Saeculares* also request that the gods grant more favorable affairs for the state: *quodque melius siet p. R. Quiritibus* (e.g. Pighi 114.92, 115.105, 117=CIL 6.32323.92).

prohibere
Aen. 3.265

The verb *prohibere* ("to keep away") appears in formal prayers for the aversion of potential trouble as early as Ennius (*scen.* 286). It also occurs in the prayer Cato prescribes for the periodic lustration of fields: *uti tu morbos visos invisosque, viduertatem vastitudinemque calamitates intemperiasque prohibessis, defendes, averruncesque* (*Rust.* 141.2).[101] There is a colloquial petition as well, *di prohibeant*, appearing in the comic playwrights and Cicero, which resembles in function the exclamation *di omen avertant* in that it either prefaces or follows a spoken reference to a potentially negative event (e.g. Plaut. *Ps.* 13-14; Ter. *Ad.* 275; Cic. *Rosc. Am.* 151).[102] The similarity of function to that of the phrase *di omen avertant* is verified by Tacitus' variant *omen . . . dii prohibeant, . . .* (*Ann.* 16.35). There is considerable variation in the construction of this petition, from the simple *quod di prohibeant* (e.g. Cic. *Fam.* 10.33.4; Ter. *And.* 568) to *verum id te quaeso ut prohibessis* (Plaut. *Aul.* 611).

Vergil's usage in the *Aeneid* recalls the context of the colloquial petition. After Celaeno has prophesied the Trojans' trials in Italy, Anchises, proclaiming sacrifices, invokes the gods, *di prohibete minas* (3.265=App. 86). Servius *auctus* compares Anchises' response to the *auspicium* with an augural prayer: *hoc per speciem auguralem quae*

---

[99] Valerius Maximus continues to tell how Scipio Africanus altered this formula when he was censor: (*sc. Africanus*) *"satis" inquit "bonae et magnae sunt; itaque precor ut eas perpetuo incolumes servent," ac protinus in publicis tabulis ad hunc modum carmen emendari iussit* (4.1.10).

[100] See Linderski (1986a) 2254 and n.423.

[101] On this prayer, see Norden (1939) 126-127.

[102] See Secknus, who finds this to be an authentically Roman expression without parallel in Menander (62).

invocatio appellatur non nulli dictum putant, invocatio autem est precatio, uti avertantur mala.[103]

ne* (*deus*) *sirit
Liv. 4.2.8, 28.28.11, 34.24.2

The earliest use of *sinere* in an apotropaic petition occurs in the archaic Arval hymn to Mars, who is asked to avert pestilence from the crops: *neve luerve Marmar sins incurrere in pleores* (Henzen CCIV=CIL 6.2104.33). The verb later appears in a colloquial exclamation, which is first attested in Plautus.[104] The speaker may address the gods in general (e.g. Plaut. *Bacch.* 468) or only the chief god, as does the youth Phaedromus from the *Curculio*:

> PA. . . . num tu pudicae quoipiam insidias locas
> aut quam pudicam oportet esse? PH. nemini;
> nec me ille sirit Iuppiter (25-27).[105]

Livy uses similar wording in three passages. For example, in a speech to the mutinous army at Sucro, Scipio prays that Jupiter not allow the ruin of Rome: *ne istuc Iuppiter Optimus Maximus sirit* (28.28.11; cf. 4.2.8, 34.24.2=App. 51, 17, 60).

Curses

One specialized type of prayer is that which asks divine powers to harm someone or to bring about a negative outcome to an event.[106] The use of self-curses in oaths is discussed separately in chapter five.

convertere
Aen. 2.191

As Sinon tells the Trojans of Calchas' prophecy of the destruction of Phrygia if the wooden horse should be damaged, he prays that the gods turn the omen of his words against the seer himself:

---

[103] Anchises' prayer is his private response to an oblative sign (cf. Liv. 7.26.4; Ov. *Met.* 9.701-703.

[104] See also Nep. *frg.* 59.

[105] Secknus identifies this as a genuine Latin formula (35-36).

[106] For a more detailed discussion of the relationship between prayer and cursing, see L. Watson 3-4.

. . . quod di prius omen in ipsum
convertant (2.190-191=App. 80).[107]

As emphasized by use of a related verb, this prayer is a variation on the colloquial formula *quod di omen avertant* (q.v.; cf. Ser. Samm. 948). The only evidence for the verb's use in any other prayer appears in a self-curse Valerius Maximus attributes to Aemilius Paulus: *precatus sum ut, si quid adversi populo Romano immineret, totum in meam domum converteretur* (5.10.2; cf. 9.1.9).

vertere
Liv. 9.1.9

Livy uses the verb *vertere* in one prayer with an entirely different function and meaning from that of propitiatory prayers such as *di bene vertant* (q.v.). This is a curse, in which the verb simply means "to turn" or "to direct." The Samnite general C. Pontius prays for divine retribution against the Romans for their refusal to make peace: *precabor, ut iras suas vertant in eos quibus non suae redditae res, non alienae accumulatae satis sint* (9.1.9). This usage of *vertere* differs from that found in other curses containing this verb. In Plautus, the intransitive propitiatory formula becomes a curse by the simple exchange of the adverb *male* for *bene* (e.g. *Curc.* 273). The only contemporary example of the formula *bene vertat* is a curse in Vergil's *Eclogues*: *quod nec vertat bene* (9.6).

---

[107] Servius (*ad Aen.* 8.484) compares this curse to that of Mezentius: *di capiti ipsius generique reservent.* Cf. Sinon's earlier reference to his doom: *quae sibi quisque timebat, | unius in miseri exitium conversa tulere* (2.130-131).

IV: Vows

Vows are prayers in which the petitioner, while making a request of a deity, also promises to give something to the deity if that request is granted. Normally the offering is made only after the granting of the request.[1]

The structure and content of public vows is well-illustrated by the *Acta* of the *Fratres Arvales,* which contain several vows for the emperor's well-being. The earliest extant vow from this collection dates to the rule of Tiberius, but the priestly college certainly made vows for Augustus as well. Furthermore, the form and wording of these imperial vows probably resembled those made earlier for the safety of the Republic.[2] One of the best-preserved vows, offered in the year 81, exemplifies the typical wording of the annual vow made on 3 January for the safety of the emperor and his household. The texts of annual vows from the years 27, 36, 38, 86, 87, 90, 91, and uncertain dates under Domitian contain similar wording but are not as well-preserved.

> Iuppiter o(ptime) m(axime), si imp(erator) Titus Caesar Vespasianus Aug(ustus) pontif(ex) max(imus) trib(unicia) potest(ate) p(ater) p(atriae) et Caesar divi f(ilius) Domitianus, quos nos sentimus dicere, vivent domusque eorum incolumis erit a(nte) d(iem) III non(as) Ian(uarias), quae proximae p(opulo) R(omano) Q(uiritium) rei p(ublicae) p(opuli) R(omani) Q(uiritium) [erunt, fuer]int, (f. *del. Pasoli*) et eum diem eosque salvos servaveris ex periculis, si qua sunt [eruntve ante] eum diem, eventumque bonum ita uti nos sentimus dicere [dederis, eosque in eo st]atu quo nunc sunt aut eo meliore servaveris, ast tu [ea ita faxsis, tunc

[1] There are a few examples of vows fulfilled before the request has been granted (Toutain 972-973). Many scholars view the *devotio* as an exception (see below s.v. *devovere*). On vows in general, see Latte 46-47; Wissowa 381-385; Marquardt 3.264-269; Toutain 971-978. For the recent and questionable view of a vow as a direct offering to a god, see Schilling (1969) 475-480, on which see Versnel (1976) 368.

[2] See Liebeschütz 65 n.2 and Weinstock 217-220 for literature. See also the discussion below concerning the formula *si res publica in eodem statu.*

> tibi nom]ine collegi fratrum Arvalium bubus au[ratis II vovemus esse futuru]m (Henzen CVII-CVIII.45-52=CIL 6.32363.45-52).

An abbreviated vow to Juno follows, which notes that the requests are identical with those to Jupiter but that the promised sacrifice will consist of two cows with gilded horns. There follow brief comments that the Brotherhood used the same wording in vows to Minerva and Public Welfare (*Salus*). Each vow invokes only one deity to whom an appropriate sacrifice is promised. Otherwise, the records indicate that the vows were identical.

The prayer is scrupulously worded. The phrases *quos nos sentimus dicere* and *uti nos sentimus dicere* are meant to protect the petitioners against any divine interpretation other than that desired. The vow states and restates the petition for well-being in a variety of different forms to ensure proper understanding on the part of the divinity. The speakers carefully allow for the possibility that the health and well-being of the emperor and his household may even improve: *[in eo st]atu quo nunc sunt aut eo meliore*. Finally, the petitioners make perfectly clear that the promise of a sacrifice is wholly conditional upon the fulfillment of their requests as specified.

Religious and Legal Obligations

As has often been noted, Roman vows bear certain similarities in both character and language to legal contracts of private law. Just as a contract obligates one or more parties to give something to or perform some action for another party, a vow also creates an obligation, which must be fulfilled under certain circumstances as defined by the maker of the vow. In both the religious and legal spheres, the verb *obligare* ("to bind") describes this relationship.[3] In one text these two spheres come together; the section *de pollicitationibus* of the Justinian digest includes a discussion of the liability of heirs for an unfulfilled vow. It begins, *si quis rem aliquam voverit, voto obligatur* (Ulp. 50.12.2). Thus at least

[3] *De voto*: e.g. Hor. *Carm.* 2.8.5; Liv. 21.21.9=App. 120; *Lex arae Aug. Narb.* CIL 12.4333=ILS 112; Macrob. *Sat.* 3.2.6; *de iure*: e.g. Plaut. *Truc.* 214; *Lex agr.* CIL 1.585.74; Cic. *Leg.* 2.41. For additional examples, see TLL s.v. *obligare* 9.2.90.55-92.79 and 93.17-19.

during the imperial period, the Romans themselves viewed certain vows as exerting a legal force. Another legal term, *damnare*, means "to condemn" or "to find guilty" someone, who is thereby obligated to suffer some penalty or make restitution. In a similar way, a person who is obligated to fulfill a vow when the request has been granted is said to be *damnatus voti/o*.[4] Finally, the verb *solvere* ("to loosen," "to pay") has an identical function when referring to both contracts and vows; the verb describes the act of fulfilling either obligation.[5]

Although there is no single type of Roman contract that corresponds exactly in all particulars to vows, both verbal and real contracts possess certain similarities to vows. As in a verbal contract (*obligatio verbis*), the maker of a vow enters the obligation by means of a verbal promise, albeit conditional. In the case of vows for the welfare of the state, this promise is a public and highly visible event, but a private vow can be silent. In another similarity to verbal contracts, a vow is unilateral in the sense that only one party is obligated. Only the human beings making the promise assume an obligation. The divinities are free to respond to the request in any way they choose and are never in a state of obligation. As in a real contract, the votive obligation becomes effective only after performance of the requested act by the deity. Should the deity fail to grant the request, the mortal speaker is under no obligation to fulfill the conditional promise.

Occasions

As abundant literary, epigraphical and archaeological evidence indicates, vows, both private and public, were a common feature of Roman religion. Private and public vows both sought protection or release from sickness and danger, in such situations as individual or communal disease, military expeditions, and journeys. Other vows sought the health and prosperity of individual, family, or state. A significant number of

---

[4] For examples of legal usage from the time of Plautus, see TLL s.v. *damno* 5.1.12.32-17.27. Examples of *damnatus voti/o* are attested from the first century B.C. (e.g. Sisenna *Hist.* 100; Nep. *Timol.* 5.3; Liv. 5.25.2; Verg. *Ecl.* 5.80).

[5] On *solutio* as the normal means of terminating a legal obligation, see A. Watson 208-212. See also VIR s.v. For *solvere* of vows, see e.g. CIL 1.1531, 1617; Verg. *Geor.* 1.436; Liv. 22.10.6=App. 123, 28.21.1. Note the standard formula found on votive offerings: *v(otum) s(olvit) l(ibens) m(erito)*.

private vows for which evidence survives were made by slaves praying for freedom.[6]

Among the *vota publica* undertaken by the Roman state were both periodic and extraordinary vows.[7] Magistrates fulfilled and renewed vows for the welfare of the state at regular intervals. These included annual vows made by consuls on the first day of their magistracy and quinquennial vows made by the censor at the close of a *lustrum.*[8] Under the emperors there were also annual vows for the safety and well-being of the emperor and his household.[9] In addition, the state might undertake extraordinary vows when it faced a dangerous situation such as a war or an epidemic (e.g. Liv. 5.19.6=App. 112, 5.21.2, 41.21.11=App. 132). Often these vows specified the period of time during which divine protection was sought (e.g. Liv. 21.62.10, 30.27.11=App. 121, 126). Commanders regularly made vows before setting out from Rome on military expeditions (*profectio*).[10]

Wars provided the occasion for the use of certain specialized types of public vows. By means of an *evocatio,* a commander sought to persuade the deity of an enemy city to abandon it to the Roman army and move to Rome where he or she would enjoy a new temple and continued worship (e.g. Liv. 5.21.3=App. 21; Macrob. *Sat.* 3.9.7-8). In a *devotio,* the commander, or a soldier appointed by him, vowed himself and/or the legions of the enemy to the *di manes* and *Tellus* in hope of victory over the enemy (e.g. Liv. 8.9.6-8=App. 115; Macrob. *Sat.* 3.9.10-11). The *devotio* differs somewhat from other types of vows in that the speaker fulfills his promise before the gods have granted the request; he charges into the enemy ranks to be killed in battle.[11]

---

[6] For discussion of vows in private cult, see De-Marchi 251-307, esp. 276-278 on occasions.

[7] On public vows, see Mommsen *StR* 1.244-246.

[8] On the annual vows, see Ov. *Pont.* 4.9.5-8 and Mommsen, *StR* 1.616, 2.412-3. On the quinquennial vows of the censor, see Val. Max. 4.1.10

[9] On the history of *vota pro salute*, see Daly 164-168.

[10] See Caes. *BCiv.* 1.6.6; Liv. 4.32.8; Fest. 176L.=173M. on *vota nuncupata.*

[11] On *evocatio,* see Basanoff and Le Gall 519-524. On *devotio,* see Janssen 357-381 and Versnel (1976) 365-410.

Audience and Function

As with other prayers, there is sometimes a dual audience and function for vows. While speakers always address requests and promises to deities, they often do so in the presence of other human beings; this is always true in the case of public vows. Although the obvious and dominant function of a vow is to secure divine assistance, some vows seem to have a secondary function. This is most obvious with vows made in the heat of battle, where the commander often seeks to persuade his soldiers of imminent supernatural aid and thus arouse them to fight harder. Livy's own wording suggests such an interpretation when he describes Romulus' attempt to stay the flight of his soldiers before the Sabines. After vowing to build a temple to Jupiter Stator, Romulus announces the god's command that the soldiers stand and fight. Livy introduces Romulus' words with the comment *veluti sensisset auditas preces* (1.12.7). The phrase *veluti* arouses a certain suspicion of Romulus' motivation. Romulus' words do have the desired effect: *restitere Romani tamquam caelesti voce iussi* (1.12.7). Again when narrating M. Atilius Regulus' vow to build a temple to the same deity during a battle with the Samnites, Livy notes that the commander makes his vow *voce clara, ita ut exaudiretur* (10.36.11=App. 118). The soldiers respond by making a renewed attempt at battle. Livy comments, *numen etiam deorum respexisse nomen Romanum visum; adeo facile inclinata res* (10.36.12). The verb *visum* casts some doubt on the agent of this turn-about. Another passage describes the effect that Cornelius Cethegus' vow to Juno has on his soldiers before they go into battle with the Insubres: *clamor sublatus compotem voti consulem se facturos* (32.30.10=App. 129). It is easy to understand what a persuasive device a vow could be in the attempt to encourage soldiers to fight, if there is any trust at all in the religious system. No doubt, it had a similarly encouraging effect on the general Roman public in other frightening situations, such as epidemics or reports of prodigies. It is unnecessary to adopt a cynical perspective toward the leaders, who need not disbelieve in the real possibility of divine aid in order to use the belief of the people to ensure success (e.g. 21.62.10, 41.21.11=App. 121, 132).

The Language of Livy and Vergil
I. Verbs of Vowing

The characteristic element of a vow is a promise to be fulfilled if and when its request has been granted. Normally, the promise contains a technical verb meaning "to vow" (*vovere* or its derivative, *devovere*), which introduces a dependent clause. Alternatively, the promise may consist of a simple future of the verb specifying the intended offering. This latter construction is more common in Vergil than in Livy. For example, Ascanius vows offerings to Jupiter if the god favors the flight of his arrow:

> Ipse tibi ad tua templa feram sollemnia dona
> et statuam ante aras aurata fronte iuvencum
> candentem pariterque caput cum matre ferentem
> (*Aen.* 9.626-628=App. 137).

devoveo

Liv. 8.9.8, 10.28

In addition to the general meaning "to consecrate," the verb *devovere* describes the vow of oneself or another to death. Its most specific usage applies to the military ritual of *devotio*, in which a commander or a soldier appointed by him vowed his own life and/or the lives of the enemy as an offering to the chthonic deities in hope of their aid against the enemy.[12] Most of Livy's uses of *devovere* refer to the *devotio* of Decius Mus or that of his son. As Decius concludes the *carmen* in which he seeks victory over the Latins, he consecrates himself and the enemy as sacrificial victims, *legiones auxiliaque hostium mecum Deis Manibus*

---

[12] The prefix *de* of *devovere* may refer to the recipients of this vow as chthonic. Janssen argues against this traditional view in part by noting that the idea of the Manes as dwelling in the underworld *per se* is late and Greek (359-360). He does not take into consideration that the Manes were from early times conceived of as dwelling in the earth and not above it (*columbaria* being a later development). Versnel (1976) distinguishes the *devotio hostium*, found in Macrobius, from the *devotio ducis*. He considers the former to be a genuine vow and the latter to be a combination of such a vow and a *consecratio* of the leader (365-410, summary 408-410). Cf. Janssen 357-381. On the *carmen* in Macrobius, see Engelbrecht 478-484. On *Aen.* 11.440-444 as a *devotio*, see Johnson 117-119. For additional examples of *devovere*, see TLL 5.881.50-882.10.

Tellurique devoveo (Liv. 8.9.8=App. 115).[13] In a similar *carmen*, Macrobius also includes the pledge *devoveo* and describes the enemy and its territory as *devotus* (*Sat.* 3.9.10-11). The regular usage of *devovere* to describe the war-time ritual of devotion identifies it as the technical verb which appeared in the actual *carmen devotionis*. *Devovere* also occurs once in a magical *defixio* (Audollent no. 129).[14]

voveo
Liv. 1.12.6, 5.21.2, 10.19.18, 22.10.2
Aen. 10.774

Not surprisingly, the technical *verbum vovendi* for most vows seems to be *voveo*. It appears in one vow found in the work of a predecessor of Livy and Vergil, a comedy by Pomponius:

> Mars, tibi facturum voveo, si umquam redierit,
> bidenti verre (*com.* 51).

Macrobius also includes it in the *carmen evocationis* attributed to Scipio Africanus: *si ita feceritis, voveo vobis templa ludosque facturum* (*Sat.* 3.9.8). Its official usage in post-republican Rome is attested by the collection of vows made by the *Fratres Arvales* (e.g. [*nom*]*ine collegi fratrum Arvalium bubus au*[*ratis II voveo esse futuru*]*m*: Henzen 100). In addition to its usage in the first person, *vovere* regularly appears in the narrative third person to describe a vow. Normally the object of the vow is expressed by a simple accusative noun (e.g. Liv. 1.12.6=App. 111), but infinitive clauses also occur (e.g. Liv. 5.19.6=App. 112).

All vows using the form *voveo* in Livy and Vergil occur in military contexts, three times in Livy and once in Vergil. Before the attack on Veii, Camillus vows a tithe of the spoils to Apollo (Liv. 5.21.2=App. 113).[15] In the very midst of fighting, Romulus and Appius Claudius vow temples to Jupiter and Bellona respectively (Liv. 1.12.6, 10.19.18=App.

---

[13] Thulin notes that some of the lines are Saturnian and considers this a word-for-word quotation (54-56). Cf. Liv. 8.11.1: *haec, etsi omnis divini humanique moris memoria abolevit nova peregrinaque omnia priscis ac patriis praeferendo, haud ab re duxi verbis quoque ipsis ut tradita nuncupataque sunt, referre*. For a somewhat more cautious view, see Skutsch (1968) 56 and notes 25, 26. (Skutsch's comparison of Livy's *ut—ita*, to earlier magical formulae is tenuous [56-57]. Note the reversal of persons.)

[14] On this derivative usage, see Versnel (1976) 396-398.

[15] This vow is followed by promises to Juno in the form of an *evocatio*. On this vow as a "mitigated *devotio*," see Versnel (1976) 382.

111, 116). During battle, Mezentius, *contemptor divum*, also makes a vow, to dedicate the armor of Aeneas to his own right hand:[16]

> dextra mihi deus et telum, quod missile libro,
> nunc adsint! voveo praedonis corpore raptis
> indutum spoliis ipsum te, Lause, tropaeum
> Aeneae (*Aen.* 10.773-776=App. 139).

Livy probably also uses the verb *vovere* in his version of the question submitted to the Roman people concerning a *ver sacrum* to expiate faults which resulted in the defeat at Trasimene: *sicut velim [vov]eamque* (*sic ed. Madvig*; 22.10.2=App. 123).

II. Requests

In vows, the most common syntax for the request is conditional clauses,[17] such as those appearing in the vows of the *Fratres Arvales*: e.g. *ast tu [ea ita faxsis, tum tibi nom]ine collegi fratrum Arvalium bubus au[ratis II vovemus esse futuru]m* (Henzen 100). The same preference appears in Livy; Appius Claudius, for example, seeks Bellona's aid during a battle with the Samnites and Etruscans thus: *Bellona, si hodie nobis victoriam duis, ast ego tibi templum voveo* (10.19.17).[18] The numerous vows in *oratio obliqua* also contain conditional clauses, as does that of Servilius Ahala after his appointment as dictator in the war against the Gauls: *si prospere evenisset ludos magnos vovit* (7.11.4=App. 114). Other constructions also appear in the requests. Livy's history contains one vow that uses an imperative (1.12.4-6=App. 111) and another in which the *verbum vovendi* introduces an adverbial purpose clause (5.21.2-3=App. 113). No request in the five vows in Vergil's *Aeneid* occurs in a dependent clause. Two are hortatory subjunctives (1.332,

---

16 In another Vergilian passage, Metabus dedicates his daughter as a servant to Diana; here *voveo* means "dedicate" (11.558=App. 107). Cf. Servius: *consecro et dico* (*ad Aen.* 10.774).

17 Cf. oaths. The concluding self-curse is conditional upon non-fulfillment of the oath. In vows, the condition is a positive one, which looks to the fulfillment of the request.

18 This usage of *ast* to introduce the apodosis of a condition is unique and perhaps intended by Livy as an archaism. On the ancient use of *ast* to introduce the protasis, see TLL 2.942.58-69.

10.774=App. 134, 139) and three are imperatives (8.73, 9.624, 10.421= App. 136, 137, 138). For example, Aeneas seeks protection and aid from the Laurentian Nymphs and the Tiber river: *accipite Aenean et tandem arcete periclis* (8.73=App. 136).

In Livy, three requests using similar vocabulary appear in more than one vow: *si prospere evenisset* (7.11.4=App. 114), *si legiones hostium fudisset* (10.42.7=App. 119),[19] and *si . . . res publica eodem stetissent statu*. Livy varies them all somewhat by means of minor expansions, substitutions, or rearrangement of words. Yet they contain technical vocabulary and retain sufficient similarity to a supposed model that they sound like genuine official formulae. Only *si res publica in eodem statu* (q.v.), however, is attested in an actual vow. The other two phrases contain collocations which relate to military matters and are sometimes attested in official documents. Rather than technical votive formulae, they appear to be Livy's inventions.

si res publica in eodem statu . . .
Liv. 21.62.10, 22.9.10, 30.2.8, 27.11, 42.28.8

In five passages, Livy describes the request of a vow with a conditional clause beginning *si res publica in eodem statu*. Varying verbs (*stare; permanere; esse*) complete the clauses. Confirmation of this phrase as a technical formula of imperial vows occurs in the *Acta Fratrum Arvalium*, in which several vows for the well-being of the emperor and his household include a phrase such as *(si) dederis eosque in eo statu quo nunc sunt aut eo meliore servaveris* (Henzen 100). Although the earliest example of such a vow dates to A.D. 27, the *Fratres Arvales* probably used similar wording in their vows for the well-being of Augustus.[20] So the language of Livy's vows at least dates to the first century. There is also some evidence to suggest that these words belong to the *vota publica* of the republic as well. Specifically, they may be part of the quinquennial vow for the welfare of the state which the censor made at the close of every *lustrum*. Two texts present versions of the prayer Scipio Africanus supposedly altered when he was censor. According to Valerius Maximus, Scipio rejected the official prayer that the gods make the state larger and more prosperous and substituted instead a

---

[19] To *prospere e.*, cf. *cetera prospera* 21.21.9; *gesta esset* 32.6.7=App. 120, 128. See also s.v. *evenire*. To *legiones f.*, cf. *hostes* 29.36.8; 35.1.9; *fusi . . . fuissent* 32.30.10=App. 124, 130, 129.

[20] On vows for the emperor, see Dio Cass. 51.19 and Henzen 90.

prayer that they preserve the present status: *precor ut eas perpetuo incolumes servent* (4.1.10). In the account found in the *Historia Augusta*, the wording of this new prayer is similar to the formula under discussion: *ut [di] scilicet in eo statu rem p. servarent* (*Max.* 17.8). The two different versions do not necessarily reflect a linguistic development; both wordings could have appeared in the original prayer. Thus the vows of the *Fratres Arvales* include prayers that Jupiter preserve (*servaveris*) the emperor and that his household be *incolumis* and *salvus*, as well as the phrase under discussion (Henzen 100).

The passages in which Livy uses the formula *si res publica in eodem statu* do not reflect the occasions of either the imperial or censorial prayers; his are expiatory (21.62.10, 22.9.10=App. 121, 122) and military (30.2.8, 30.27.11, 42.28.8=App. 125, 126, 133). This could indicate either that the formula was common and non-specific or that Livy himself has broadened its application.

voti reus
Aen. 5.237

During the funeral games for Anchises, Cloanthus vows sacrifices to the gods of the sea as his competitor threatens to overtake him near the finish line:

> di, quibus imperium est pelagi, quorum aequora curro,
> vobis laetus ego hoc candentem in litore taurum
> constituam ante aras voti reus, extaque salsos
> proiciam in fluctus et vina liquentia fundam
> (*Aen.* 5.235-238=App. 135).

The exact meaning of the rare phrase *voti reus*, here attested for the first time,[21] is the subject of some disagreement. While lexicographers hold that the maker of a vow is *reus* only after his request has been granted, Toutain and other scholars hold the opposite point of view—that a person is *reus* from the moment he makes his conditional promise until the god has either fulfilled the request or failed to do so.[22]

The obvious starting point for evaluation of these views is the Vergilian passage itself. Vergil's usage is unambiguous; the *voti reus*

---

[21] Later occurences appear in Petr. *frg* 27.12; Stat. *Theb.* 6.198; CIL 8.8259.

[22] For the former view, see OLD, Lewis and Short, and Forcellini s.v. *reus*. For the latter, see Wachsmuth 1357; Wissowa 382; Toutain 974.

here is one whose prayer has been granted. Cloanthus speaks of his desired future status as victor. At that time he will place (*constituam*) a sacrificial victim before the altars of the gods who have granted him victory. It is to the future victor, Cloanthus, that the phrase *voti reus* applies. Furthermore, Vergil describes Cloanthus as *laetus*, another reference to him as victor and thus one whose prayer has been granted. *Laetus* is here, as in some imperial inscriptions, synonymous with the term *libens*, which appears so frequently on votive offerings of thanksgiving.[23] These two terms, *voti reus* and *laetus*, express in an abbreviated manner the request which would normally be contained in a separate clause; Cloanthus is asking the gods of the sea for victory.

The words of imperial commentators on Vergil also comprise important evidence in the discussion of the phrase *voti reus*. Servius' comments on the phrase *voti reus* in this passage indicate that the term applies to a person whose prayers have been answered. He equates the phrase with *voti debitor* and explains *inde est damnabis tu quoque votis, quasi 'reos facies.'* The reference here is to *Eclogues* 5.80: *damnabis tu quoque votis* "you will condemn people to pay vows by granting their requests." Commenting on this phrase from the *Eclogues*, Servius writes, *id est cum deus praestare aliqua hominibus coeperis, obnoxios tibi eos facies ad vota solvenda, quae ante quam solvantur, obligatos et quasi damnatos homines retinent* "that is, when as a god you undertake to perform something for human beings, you will make them obligated to discharge their vows, which keep them obligated and as it were, almost condemned men, until they are discharged" (*ad Ecl.* 5.80). Here the word *obnoxios* is equivalent to *reos*, which also occurs with the verb *facere* in the same passage. Commenting on *Aeneid* 4.699, which contains the verb *damnare*, Servius *auctus* refers to the passage from the *Eclogues* and uses the phrase *reus voti*: *et cum vota suscipimus, rei voti dicimur donec consequamur beneficium et donec condemnemur, id est, promissa solvamus, ut damnabis tu quoque votis* "when we undertake vows, we are said to be *rei voti* until we obtain the favor and until we are freed, that is, until we discharge our promises, as *damnabis tu quoque votis*."[24] Macrobius' comments on the phrase *voti reus* repeat the interpretation found in the last comment of Servius *auctus*: *haec vox propria sacrorum*

[23] N.B. the standard formula: *v(otum) s(olvit) l(ibens) m(erito)*. Cf. imperial inscriptions which contain the two synonyms (ILS 3824, 3929, 7310).

[24] On the meaning "to free from a vow" for *condemnare*, see Nonius 277.

est, ut reus vocetur, qui suscepto voto se numinibus obligat, damnatus autem qui promissa vota iam soluit "this is a religious technical term, as whoever binds himself to the gods by undertaking a vow is called *reus*; however, whoever has already discharged his vows is called *damnatus*" (*Sat.* 3.2.6).[25] These two explications extend the initial temporal boundary of the word *reus* backwards to the time when a vow is taken, but retain the terminal boundary at the time when the vow is discharged.[26] A person is *reus* until he has paid his vow; this categorization does not end at the time of the divine response.

Although Vergil has avoided the more common expressions, *voti compos* or *voti/o damnatus*, his diction does reveal his appreciation of the legalistic character of Roman vows.[27] In choosing the noun *reus*, he has selected a common legal term frequently designating the defendant. The noun could also, however, refer to one found guilty. Either meaning seems appropriate for a person from whom a god seeks, as it were, to collect a debt.

III. Cautionary Formulae

As was generally the case with their religious rituals, the Romans were very scrupulous about specifying all details involved in vows, especially the conditions under which a vow was considered to be fulfilled.[28] Thus at the end of a vow, there might be a list of situations that could conceivably occur and that would invalidate the offering intended to discharge the vow. The petitioner would declare that despite such occurrences, the offering would, nevertheless, be considered valid. Both the words *recte* and *probe*, which occur in these formulae, are technical terms of religious language meaning "properly" or "in accordance with

[25] Note that Lewis and Short emends the passage to read *vota non soluit*. On *damnare* meaning "to discharge vows," see Nonius 277.

[26] Stat. *Theb*. 6.198 seems to refer to the time the vow is undertaken.

[27] E.g. *voti compos*: Liv. 7.40.5=App. 31, Suet. *Aug*. 28.2, CIL 1.2520.18 and *Act. Frat. Arv*. (e.g. Henzen 110-111 for A.D. 86, 87, and 90). For additional examples of both phrases, see TLL s.v. *compos*; *damno*; cf. *condemno*.

[28] On these provisos, see Guittard (1981) 9-20 and Nock 481-492.

ius divinum."[29] It is not that the Romans placed themselves on a level with the gods and could thus dictate the terms of their relationship. This concern for exactitude in their dealings with the gods shows a certain fear before their power and apparent unpredictability. As with any other religious ritual, incorrect performance of a vow in any detail invalidated the entire attempt and necessitated the repetition of part or all of the ritual. This is clearly illustrated in Livy's account of the *ver sacrum* (22.9.7-22.10.6=App. 122). In response to the disastrous defeat at Trasimene, the dictator Fabius Maximus set about to re-establish the *pax deum*, without which the Romans could not hope for the successful outcome of any military endeavor. Therefore he ordered the consultation of the Sibylline Books to determine *quaeque piacula irae deum essent* (22.9.7). The Books revealed that the fault was in an earlier vow to Mars on account of the war: *id non rite factum de integro atque amplius faciundum esse* (22.9.9). In expiation for that fault, the state would have to undertake new vows, including the unusual vow of a *ver sacrum*, the sacrifice of all herd animals born in one spring season. In light of this background, it is not surprising that the vow of the *ver sacrum* would be very carefully worded to avoid as many potential faults as possible:[30]

> Qui faciet, quando volet quaque lege volet facito; quo modo faxit probe factum esto. Si id moritur quod fieri oportebit, profanum esto, neque scelus esto. Si quis rumpet occidetve insciens, ne fraus esto. Si quis clepsit, ne populo scelus esto neve cui cleptum erit. Si atro die faxit insciens, probe factum esto. Si nocte sive luce, si servus sive liber faxit, probe factum esto. Si antidea ac senatus populusque iusserit fieri faxitur, eo populus solutus liber esto (22.10.4-6=App. 123).

In addition to illustrating the religious scrupulosity of the Romans, this list also demonstrates once again the similarities in both character and language between Roman religion and law. The collocations *probe facere* and especially *recte facere* belong to the language of law as well as

[29] Cf. similar phrases such as *fas esto* or *ius esto*, which also appear in *leges ararum*; see Nock 482-483.

[30] Only one other vow in Livy includes such a list of provisos; it also occurs at a time of war (36.2.5=App. 131).

religion. The use of the future imperative *esto* is also typical of both legal and religious documents.[31]

probe factum esto
Liv. 22.10.4, 6 (*bis*)

The question put to the Roman people concerning the vow of a *ver sacrum* repeats the phrase *probe factum esto* three times (22.10.4, 6=App. 123). The conditions allow the offering to be performed by any method, by day or night, by slave or free man. They also cover the event of an unwitting offering on a *dies ater*. None of these possible violations is to disqualify the offering; it is to be considered *probe factum*. There are no other examples of this phrase in the context of vows. Livy does use a similar phrase when describing the ritual of *devotio*: *si is homo qui devotus est moritur, probe factum videri* (8.10.12). Evidence for the authenticity of this phrase in technical religious language derives from two imperial *leges ararum*. The *lex Narbonensis* reads, *sive quis hostia sacrum faxit, qui magmentum nec protollat, idcirco tamen probe factum esto*; it probably preserves the wording of the *lex arae* of the temple of Diana on the Aventine (CIL 12.4333=ILS 112). The *Lex Salonitana* contains almost the same words (CIL 3.1933=ILS 4907).

recte factum esto
Liv. 36.2.5

Recte factum esto is synonomous with the formula just examined, but seems to be older and certainly more common.[32] In the prayer to precede the thinning of a grove, Cato uses an almost identical formula asking the deity to accept the expiatory sacrifice, *sive ego sive quis iussu meo fecerit, uti id recte factum siet* (*Rust.* 139).[33] Macrobius also includes this *formula* at the end of Scipio's *carmen devotionis*: *tunc quisquis votum hoc faxit ubiubi faxit recte factum esto ovibus atris tribus* (*Sat.* 3.9.11).[34] *Recte* appears frequently in religious contexts to indicate that a ritual act is performed properly. Thus Varro writes, *quod enim fit*

---

[31] On the use of the future imperative, see Norden (1939) 40.

[32] In non-religious contexts, the collocation *recte facere* is far more frequent than *probe facere* (e.g. Cic. *Quinct.* 8.31 *et saepius*). For continued preference of *recte*, compare entries for *probe* and *recte* in the VIR.

[33] A similar idea is expressed in an ancient votive inscription from Falerii: *Menerva sacru Lr. Cotena Lr. f. pretod de zenatuo sententiad vootum dedet; cuando datu rected cuncaptum* (i.e. *quando datur recte conceptum*) (CIL 1.365=ILS 3124).

[34] On the unusual construction, see Versnel (1976) 386-387 with n.59.

rite, id ratum ac rectum est; ab eo Accius: "recte perfectis sacris" (*Ling.* 7.88). Livy uses it thus too: *sacrificia . . . quae absente se recte fieri non possent* (42.32.2). Two republican laws attest its legal usage as well to mean that an act is done in accordance with the law (CIL 1.583.30, 65, 67, 68; 1.594.128). Livy uses this formula in one vow. When the consul M'. Acilius Glabrio vows games and offerings to Jupiter prior to the war with Antiochus, he concludes with these words: *quisquis magistratus eos ludos quando ubique faxit, hi ludi recte facti donaque data recte sunto* (36.2.5=App. 131).

V: Oaths

The basic function of an oath is affirmation of the speaker's words. In order to accomplish this, the speaker either explicitly or implicitly makes two requests of the divinities by whom he or she swears; thus an oath is a type of petitionary prayer.[1] The primary request is a self-curse that the speaker suffer some punishment if the words are intentionally false. The secondary request, that the divinities be present as witnesses to the speaker's words, is a logical accompaniment to the first idea. Punishment is dependent upon proof that perjury has been committed; without a witness to the original statement, there can be no proof of perjury.

The only epigraphical evidence for a Latin oath that is preserved in its entirety is the oath of allegiance taken by the people of Aritium in Spain to the emperor Caligula upon his accession in A.D. 37. Despite its date and imperial subject, this oath contains certain elements and formulae characteristic of earlier Republican oaths as well.

> Iusiurandum Aritiensium.
>
> Ex mei animi sententia, ut ego iis inimicus ero, quos C. Caesari Germanico inimicos esse cognovero, et si quis periculum ei salutiq(ue) eius in[f]ert in[f]er[e]tque, armis bello internecivo terra mariq(ue) persequi non desinam, quoad poenas ei persolverit, neq(ue) me neque liberos meos eius salute cariores habebo, eosq(ue) qui in eum hostili animo fuerint, mihi hostes esse ducam; si s[cie]ns fa[ll]o fefellerove, tum me liberosq(ue) meos Iuppiter Optimus Maximus ac Divus Augustus ceteriq(ue) omnes di immortales expertem patria incolumitate fortunisque omnibus faxint (CIL 2.172=ILS 190).[2]

---

[1] There are many literary oaths in which a speaker names only human witnesses, objects, or none at all; see Hirzel 15-22. The following discussion examines only those oaths naming divine witnesses. On early Roman concepts concerning oaths and punishment, see Kaser 21. On self-curses, see Hirzel 8. On gods as witnesses, see Hirzel, esp. 11, 23-27. Cf. Cic. *Off.* 3.104: *quasi deo teste*.

[2] Cf. the fragmentary Latin oath of Sestinum (CIL 11.5998a and Herrmann appendix 1 with other loyalty oaths).

The oath of the Aritiensians contains two phrases referring to the mental state of the speaker. The oath begins with the words *ex mei animi sententia*, indicating that the speaker fully understands the promises which follow. After the promises, the speaker again refers to personal knowledge, this time in connection with any violation of the oath: *si sciens fallo*. These phrases recall the cautionary formulae of vows and reflect the same legalistic approach in dealings with the gods. The legal concept that distinguishes between an action and the intent of its agent dates back as early as the regal period, as we know from Cicero's references to a law from the Twelve Tables: *si telum manu fugit magis quam iecit* "if a weapon slips from the hand rather than someone throwing it" (*Top*. 64; *sim. Tull.* 51 and *Off.* 3. 108).

The oath of allegiance to Caligula also illustrates well the original, fundamental concept of all oaths: that the person knowingly forswearing an oath is liable to divine punishment. Thus the speaker calls upon Jupiter, the chief god of oaths, and other gods to punish any violation of the promise.[3] The curse of exile is a common one, which also occurs in the traditional oath by Jupiter. In that oath, the speaker symbolically casts a stone and prays that the god will similarly cast him out of his homeland if he violates his oath: *si sciens fallo, tum me Dispiter salva urbe arceque bonis eiciat, ut ego hunc lapidem* (Paul. Fest. 102L.=115M.).[4]

The oath of the Aritiensians makes no reference to gods except in the curse; it lacks the introductory invocation characteristic of formal Roman prayers. Yet in an oath it was important for the gods to be attentive as witnesses to the speaker's promise as well at to the curse; otherwise, they would not be able to judge whether the speaker kept his

---

[3] Concerning punishment for breaking an oath, see Wissowa 388 with notes. On Jupiter as the primary god of oaths, see Wissowa 118. The *penates* are also frequently invoked, e.g. the Republican *Lex Bantina* (CIL 1.582.24).

[4] Cf. Polyb. 3.25.6: τὸν δ' ὅρκον ὀμνύειν ἔδει τοιοῦτον . . . 'Ρωμαίους δὲ Δία λίθον κατά τι παλαιὸν ἔθος . . . ἔστι δὲ τὸ Δία λίθον τοιοῦτον· λαβὼν εἰς τὴν χεῖρα λίθον ὁ ποιούμενος τὰ ὅρκια περὶ τῶν συνθηκῶν, ἐπειδὰν ὀμόσῃ δημοσίᾳ πίστει, λέγει τάδε· εὐορκοῦντι μέν μοι εἴη τἀγαθά· εἰ δ' ἄλλως διανοηθείην τι ἢ πράξαιμι, πάντων τῶν ἄλλων σῳζομένων ἐν ταῖς ἰδίαις πατρίσιν, ἐν τοῖς ἰδίοις νόμοις, ἐπὶ τῶν ἰδίων βίων, ἱερῶν, τάφων, ἐγὼ μόνος ἐκπέσοιμι οὕτως ὡς ὅδε λίθος νῦν. καὶ ταῦτ' εἰπὼν ῥίπτει τὸν λίθον ἐκ τῆς χειρός; Plut. *Sull.* 10.4: ὁ δὲ ἀναβὰς εἰς τὸ Καπιτώλιον ἔχων ἐν τῇ χειρὶ λίθον ὤμνυεν, εἶτα ἐπαρασάμενος ἑαυτῷ μὴ φυλάττοντι τὴν πρὸς ἐκεῖνον εὔνοιαν ἐκπεσεῖν τῆς πόλεως, ὥσπερ ὁ λίθος διὰ τῆς χειρός, κατέβαλε χαμᾶζε τὸν λίθον οὐκ ὀλίγων παρόντων. On this much discussed oath, see Walbank, vol. 1, 351-353 with literature and Wissowa 552-553.

promise or not. Thus the omission of the invocation in this document invites some comment. It is noteworthy that the Greek text of a similar loyalty oath to Caligula by the people of Assos in the Troad does begin with a list of the gods by whom the speakers swear: ὄμνυμεν Δία Σωτῆρα καὶ θεὸν Καίσαρα Σεβαστὸν καὶ τὴν πάτριον . . . (IG Rom. 4.251=Syll. Graec. 797=Herrman app. 1). This element appears in other Greek loyalty oaths as well.[5] The lack of an invocation in the Latin inscription does not necessarily mean that there was none spoken. It may have been so standard that it was felt unnecessary to transcribe the words, perhaps *iuro per Iovem Optimum Maximum ac Divum Augustum deosque omnes immortales.*[6]

The evidence from literary sources supports the idea that formal Latin oaths, when expressed in their fullest form, began with some sort of invocation of divine witnesses.[7] The best evidence may be that provided by Livy's accounts of oaths used at the solemnization of a treaty (1.24.7=App. 141) and in the ritual for demanding satisfaction for wrongs done by the people of another state (1.32.6=App. 142). Both oaths begin with an invocation of Jupiter. Following the statements to be witnessed, the oaths conclude with curses, which again name the chief god of oaths. Other evidence for the invocation of witnesses is derived from literary oaths in non-official situations. One Plautine parody contains all of the elements found in Livy's oaths. In the *Rudens*, the fisherman Gripus has the pimp Labrax swear that the latter will give Gripus whatever money is found in a lost trunk. Gripus dictates the words, invoking Venus both as witness and as the agent of divine punishment:

GR. Venus Cyrenensis, testem te testor mihi,
si vidulum illum quem ego in novi perdidi
cum auro atque argento salvom investigavero

[5] E.g. the oaths from Phazimon in Paphlagonia, 3 B.C. (Cumont 27) and Palaipaphos on Cyprus, A.D. 14 (Mitford 75). Herrmann includes both of these oaths in appendices 1 and 2. On the traditional god-list in treaties of the ancient Middle East, see Barré. A comparative study of the structure of various types of oaths among ancient civilizations including the Romans is inappropriate here but deserves future attention. Herrmann focuses on *leges* rather than on oath formulae, while Seidl emphasizes occasions and guarantors.

[6] Cf. the abbreviated report of oaths in Roman Egypt, e.g. ὀμνύω τὸν Ῥωμαίοις ἔθιμον ὅρκον (see Seidl 33-34 and 134).

[7] See Täubler 132 on the existence of an invocation at the beginning of Roman treaties.

isque in potestatem meam pervenerit,
tum ego huic Gripo, inquito et me tangito—
LA. tum ego huic Gripo (dico, Venus, ut tu audias)
talentum argenti magnum continuo dabo.
GR. si fraudassis, dic ut te in quaestu tuo
Venus eradicet, caput atque aetatem tuam
(*Rud.* 1338-1346).

Since the humor of a parody derives from the tension between the simultaneous similarity and dissimilarity of the parody and the object of parody, it is likely that this Plautine passage closely reflects the structure of an actual oath. Finally, literary evidence also includes numerous oaths which contain only invocations. Cicero, for example, on several occasions invokes the gods as witnesses to some statement: *vos, di patrii ac penates . . . testor integro me animo ac libero P. Sullae causam defendere, . . .* (*Sull.* 86).

Brief oaths may only allude to the presence of divine witnesses by means of a phrase such as *per superos* (*Aen.* 6.458-459=App. 170). The relationship of these abbreviated oaths to those directly invoking witnesses is demonstrated by the frequent association of words meaning "to call to witness" (e.g. *testor*) with oaths of both the abbreviated and complete types. The ultimate reduction of oaths is seen in asseverative expressions such as *hercule*, which probably originated from a petition similar to *ita me Hercules amet.*

In literary prayers, the curse is commonly omitted. None of the oaths in Vergil's *Aeneid* makes reference to punishment and only a few of those in Livy do.[8] The same tendency is true in other authors. In fact, oaths in comedy frequently illustrate a quite different pattern: they request that the gods be propitious, for example, *ita me Iuppiter, Iuno . . . dique omnes ament ut . . .* (Plaut. *Bacch.* 892-896). If completely expressed, however, the thought here would be: if I speak truthfully, may the gods be propitious to me; if I do not, may they punish me.[9] In the majority of cases, the author probably gave little or no thought to the

[8] Punishment is implied at *Aen.* 9.207-209=App. 171. Seidl notes that one-third of imperial oaths in Roman Egypt lack a *sanktionsformel* (120).

[9] Cf. the Greek formula εὐορκοῦντι μέν μοι εὖ εἴη, ἐπιορκοῦντι δὲ τὰ ἐναντία (e.g. POxy. 253, 1453).

original meaning of the oaths which he used. Nevertheless, the linguistic patterns show that the original form did include both the concepts of witness and of punishment.

Occasions

The ancient Romans used formal public oaths in a variety of situations: legal, military, administrative, and judicial. One of the oldest occasions for a public oath was the ratification of a treaty between two communities. For this task, the Romans and other Italian peoples had a religious brotherhood, the *Fetiales*, which performed the actual religious ceremony, wherein a sacrifice accompanied an oath to abide by the terms of the treaty.[10] A more common oath, termed *sacramentum*, was that sworn by conscripts when they entered the army.[11] Their basic pledge was to fight under the direction of their commander and not to desert for whatever length of time the state kept them in service. During the Republic, this oath was taken in the name of the commanding officer; under the Empire, in the name of the emperor. In the last century of the Republic, soldiers sometimes took oaths of allegiance to a private leader for a particular war. Such military oaths may have furnished a model for later civilian oaths of allegiance to the emperor and his family as illustrated by the above oath of Aritium to Caligula.[12] A related oath, which also had its beginnings at the close of the Republic, was the oath to support the legal acts (*acta*) of the emperor and his predecessors, an oath which Senators and magistrates swore annually on New Year's Day. According to Appian, magistrates first took such an oath in 45 B.C. in honor of Julius Caesar (App. *BCiv*. 2.106).[13] The traditional oath taken by magistrates during the Republic was not to support the acts of an individual, but to support the laws of the state (e.g. *Lex Salpens*. 26, CIL 2.1963; *Lex Malac*. 59, CIL 2.1964).[14] As part of his inauguration, every

---

[10] On the Fetial brotherhood, see Harris 166-171, Bayet 29ff., Rich *passim*, and Wissowa 550-554. On treaties, see Taübler 132ff. and Mommsen *StR* 1.246-257.

[11] On the *sacramentum*, see Campbell 19-32 and RE 1A.1667-1668. For references in Livy, see 2.45.14=App. 146, 8.34.10, 28.29.12.

[12] On *genera militiae* and corresponding oaths, see Linderski (1984) 77-80. See also Campbell 19-23 on oaths of allegiance to individual leaders. On loyalty oaths to the emperor, see Herrmann 50-121, Weinstock 223-227, Syme (1939) 284-293.

[13] On oaths *in acta*, Weinstock 222-223 and Sherwin-White 611-612.

[14] On oaths *in leges*, see RE 10.1257.

magistrate swore such an oath publicly and delivered a written version to the quaestors. Upon his departure from office, every magistrate likewise took an oath that he had done nothing contrary to the laws of the state. During the course of a year, the enactment of a new law might require Senators and magistrates to swear to uphold that specific law.

In addition to these oaths related to military and civic life, oaths also played a part in the judicial system of Rome. The most common form of early legal action (*legis actio*) required the deposit of a sum of money by both parties to a suit; the loser's money was forfeited to the state. The term for this deposit, *sacramentum*, has led many scholars to suggest that it was derived from an earlier proceeding in which the *sacramentum*, as its name implies, was actually an oath. In that situation the party whose oath was proven false was obligated to make an expiatory offering to the gods. In the later formulary process, oaths could be used in various ways. The challenge by one party that the other swear an oath to the validity of a claim was one means of bringing an early termination to an action brought before a magistrate (*in iure*) when the challenged party refused to swear. In some situations, however, the refusal to swear an oath did not result in the loss of the action; this was true, for example, when such a challenge was made once the proceedings were before the judge (*apud iudicem*). At that stage oaths could also serve to establish the value of the thing at issue.[15]

Audience

Oaths regularly address a dual audience—human and divine.[16] By definition, there is always a divine audience, who also function as witnesses. In addition, there is almost always a human audience. Often the invocation of witnesses summons both human and divine (e.g. *deos hominesque* Liv. 3.72.1=App. 150). Sometimes the speaker explicitly addresses only one of the audiences, the human or the divine. For example, Brutus, standing over the lifeless body of Lucretia, directly addresses the gods and makes them witnesses of his promise never to allow another

---

[15] See Kaser 197-200 with references on *parteieneid*.

[16] Here I distinguish between witnesses in the technical sense of those specifically named as witnesses to the speaker's words and an audience which hears those words but is not necessarily invoked as witnesses. The term audience here subsumes the term witness.

king at Rome: *vos . . . di testes facio* (Liv. 1.59.1=App. 144).[17] A large human audience, however, also hears and responds to these words. Conversely, the speaker may address a human audience and merely allude to the divine. Thus, the Greek Achaemenides begs the Trojans *per superos* to carry him away from the island of the Cyclops (*Aen.* 3.599=App. 167).

Function

Oaths frequently have multiple functions. First, they have the social function of guaranteeing the trustworthiness of the speaker to other human beings and, in the event of perjury, making a provision for punishment. Second, oaths have the rhetorical function of persuasion, especially of the human audience. In fact, people who invoke divine witnesses often appears to be far more interested in the effect of their words on the human audience than on the divine. Thus, the primary objective of Achaemenides' speech to the Trojans is to persuade the latter to take certain actions (*Aen.* 3.598-600=App. 167). His reference to the gods is a means towards that goal. Similarly, when Horatius Cocles invokes gods and men to witness the futility of flight, his primary objective is to persuade the soldiers to stand firm and fight (2.10.3=App. 145).

In both of these passages, oaths function to emphasize or lend weight to the speaker's words. Achaemenides, for example, wants to persuade the Trojans of the importance and urgency of his request. This emphatic function is a development from the more basic function of affirming the speaker's truthfulness. Any request or statement is effective only to the degree that the listener perceives it to be true. Thus, in order to end the flight of his fellow soldiers, Horatius must persuade them that his statement concerning the futility of flight is true. It is for this reason that he invokes witnesses. The force of such an invocation stems from the idea that the speaker would be held liable to punishment if he should speak falsely. Similarly, a treaty cannot be effective unless the two parties place credence in the truthfulness of the promises spoken by each side. Therefore, the ritual for solemnization of a treaty includes the invocation of divine and human witnesses and, as an additional affirmation, a curse upon the state that swears falsely (Liv. 1.24.7-8=App. 141).

---

[17] Cf. the subsequent oath of the citizens not to allow a king to rule among them (2.1.9). On this oath, see Mommsen *StR* 2.16, 3.362.

A general correlation exists between the fullness of expression (i.e. inclusion of direct invocations and curses) and the concern with truthfulness. For example, the fetial formula for the solemnization of a treaty illustrates a full oath which primarily functions to affirm the speaker's truthfulness. The fetial formula of *rerum repetitio* is clearly intended to affirm the speaker's truthfulness: *verbis fides sit* (Liv. 1.32.6-8=App. 142). Achaemenides' oath (*per superos*), however, which does not directly address the gods and makes no mention of punishment, is used more for emphasis than for affirmation. This emphatic function is primary for almost all oaths accompanying requests. Finally, elliptical expressions such as *hercule* are scarcely more than colloquial interjections intended to give emphasis to the speaker's words; they mean little more than "indeed," for example, *vobis mehercule, Martiis viris, cavenda ac fugienda quam primum amoenitas est Asiae* (Liv. 38.17.18=App. 165).[18]

The Language of Livy and Vergil
I. Verbs of Oath-taking

The reference to witnesses occurs either directly or indirectly. When the speaker invokes the gods directly, he or she uses a first person singular verb meaning "I summon a witness" or "I make someone a witness" (*testor, testem facio,* and *obtestor*). The second person imperative and subjunctive forms of the verb *audire*, which appear in Livy's versions of the fetial formulae serve a similar function but shift the perspective from the act of the speaker to that of the witness. Similarly, Vergil's phrase *esto testis* expresses the invocation in terms of the deity's act. When the speaker invokes the gods indirectly, he or she refers to witnesses by means of a prepositional phrase with *per* and then proceeds with a verb describing the nature of the communication. Thus certain verbs, including *iuro*, *oro*, and words with similar meanings (sometimes *testor*), frequently appear with a phrase such as *per deos*. In addition to their function as *verba testandi*, these verbs sometimes serve as *verba precandi*.

An examination of the grammatical contructions used in association with oaths in Livy and Vergil reveals certain patterns. The most common context for oaths is a statement. These may concern either facts, as perceived by the speaker (e.g. Horatius' statement of the futility of flight),

[18] Steinwenter categorizes oaths as either *assertorische* or *promissorische*. He distinguishes *Beteuerungen* and pleas from oaths (10.1.1253-1254).

or promises (e.g. the fetial formula for solemnization of a treaty).[19] In Livy, statements concerning facts contain a verb of oath-taking followed by an accusative with infinitive construction. One exception contains an indirect question (1.22.7=App. 140). The Vergilian constructions include two accusatives with infinitive (4.492, 12.581=App. 168, 175). Oaths accompanying requests employ either a *verbum testandi* or a *verbum precandi*. In Livy these verbs introduce noun jussive clauses (e.g. 6.14.5, 21.10.3=App. 156). In Vergil the requests are sometimes expressed by imperatives (e.g. 3.601, 9.257=App. 167, 172).

audire

Liv. 1.24.7, 1.32.6, 9, 3.25.8, 8.5.8
Aen. 8.573, 12.200

No Latin prose, either literary or documentary, other than Livy's history, uses the verb *audire* in a prayer. There are, however, a few references to the possibility of gods hearing prayers, which could indicate that this was a common, or at least debated, notion, for example, Labrax' words *dico, Venus, ut tu audias* (Plaut. *Rud.* 1343; see also Cic. *Dom.* 104; Quint. *Decl.* 329).

The most intriguing observation concerning the verb, "to listen," is its association with treaty oaths in the works of Livy and Vergil and in Hittite treaties.[20] All uses of *audire* in prayers in Livy either appear in fetial formulae or allude to them. As Ogilvie notes in his commentary, the formulae are presumably "an archaizing reconstruction" of the second century. "Such formulae will first have been published in manuals of constitutional procedure and then been incorporated by annalists into their histories" (110). When solemnizing the treaty with the Albans, the *pater patratus*, with a threefold repetition of *audi*, calls on Jupiter, the Alban *pater patratus*, and the Alban people to hear his oath (1.24.7-8=App. 141). Another threefold invocation begins the fetial formula demanding satisfaction for treaty violations: *audi, Juppiter, . . . audite fines . . . audiat fas* (1.32.6=App. 142).[21] Again, in the declaration of war, the formula begins by requesting the attention of various gods: *audi, Iuppiter, et tu Iane Quirine, dique omnes caelestes, vosque terrestres*

---

[19] See Hirzel 2-7, esp. 2 n.2, on types of oaths.

[20] See Barré 34, 156 n.162 for citations of five examples in Hittite treaties.

[21] On *audiat fas*, see Ogilvie (1965) 130-131. See Thulin 63-64 for a reconstruction of this formula.

vosque inferni, audite (1.32.9=App. 143). In a literary variation on these formulae, an envoy calls the gods to witness treaty violations by the Aequians thus: *et haec . . . sacrata quercus* (*sc. Iuppiter*) *et quidquid deorum est audiant foedus a vobis ruptum* (3.25.8=App. 148). Finally, addressing the Senate, T. Manlius dramatizes the requests of the Latin envoy: *audi Iuppiter haec scelera . . . audite Ius Fasque* (8.5.8=App. 153). Like Livy's history, the *Aeneid* includes an oath for the solemnization of a peace treaty which begins by invoking Jupiter. Latinus seeks a divine witness for the treaty between Trojans and Latins: *audiat haec genitor qui foedera fulmine sancit* (12.200=App. 174). In Plautus' parody of an oath, Labrax makes reference to the importance of Venus' hearing the words of the oath between two parties: *dico, Venus, ut tu audias* (*Rud.* 1343). Cicero's brief colloquial appeal to his friend Trebatius, *audi, testa mi*, may give further evidence for the association of the verb *audire* with oaths (*Fam.* 7.13.1).

In other poetic prayers, *audire* appears in a wide variety of contexts. The earliest appearance in a prayer occurs in Ennius' version of a *carmen devotionis*, in which the petitioner asks that the gods listen to his vow:

> . . . Divi hoc audite parumper,
> ut pro Romano populo prognariter armis
> certando prudens animam de corpore mitto
> (*Ann.* 208-210).[22]

Horace also uses this verb in the hymn which he composed for performance at the Augustan Secular Games. The chorus of girls and boys prays for the attention of Apollo and Luna:

> condito mitis placidusque telo
> supplices audi pueros, Apollo;
> siderum regina bicornis, audi,
> Luna, puellas: . . . (*Carm. Saec.* 33-36).[23]

Besides the usage of *audire* to request that the gods listen to a prayer, the verb often carries the force of *exaudire*, "to grant."[24] When

---

[22] On this hymn, see Skutsch (1968) 54-61.

[23] Cf. other poetic prayers: *huc huc adventate, meas audite querellas* (Catull. 64.195) and *huc ades atque audi placidus, Neptune, precantem* (Ov. *Met.* 8.598). For additional examples, see TLL 2.1271.46-72. For examples of Greek *verba audiendi*, see Ausfeld 516-517 (e.g. κλῦθι: *Il.* 1.37; ἄκουε: Soph. *El.* 643).

[24] Ernout-Meillet[4] s.v. *audio* and TLL 2.1289.83-1290.24.

Evander prays for Pallas' safety in battle (*miserescite regis et patrias audite preces*: *Aen.* 8.573-574=App. 99), he asks for the fulfillment of his prayer. Dido's curse of Aeneas may represent an ascending tricolon ("hear," "attend to," and "grant"):

> accipite haec, meritumque malis advertite numen
> et nostras audite preces (*Aen.* 4.611-12=App. 90).[25]

This meaning is more obvious when *audire* is used formulaically, as in several Vergilian passages, to indicate that the god has granted the prayer just spoken, for example:

> talibus orantem dictis . . .
> audiit Omnipotens (*Aen.* 4.219-220).

This narrative usage also appears in other authors of both prose and poetry (Liv. 6.23.12; Cic. *Pis.* 43).

esto testis

Aen. 12.176

Preceding the duel between Aeneas and Turnus, Aeneas and Latinus take formal oaths binding themselves and their peoples to the outcome of this duel and fixing the provisions of the treaty between these two peoples. Aeneas begins his oath with a lengthy catalogue of divine witnesses introduced by the words *esto nunc Sol testis* (12.176=App. 173). The phrase *esto testis* is unique. *Testis* does appear frequently with the verb *esse* but when it does the verb is in the indicative mood and refers to human witnesses. It is likely that the phrase is derived from an oath in Homer's *Iliad*: μάρτυροι ἔστε (3.280; cf. *sim. Il.* 1.338, 2.302.) The Greek oath similarly precedes a duel, that between Paris and Menelaus. Among Vergil's successors, Ovid perhaps shows the influence of this formulation: *testes estote* and *testis adesto* (*Fast.* 3.707; *Met.* 2.45).

iuro

Liv. 1.59.1

Aen. 6.458, 12.197

The verb *iurare* is a common term throughout Latin literature and legal inscriptions.[26] It appears both in the first person within the actual

---

[25] TLL 2.1290.4 classifies *audire* here as *exaudire*, but Conington-Nettleship understands it as *audire*. On *accipere* as *audire*, see Servius on 6.66 and Servius *auctus* on 4.611.

body of oaths as well as in the third person in a narrative context. Unlike the verbs and verb phrases formed on the root *test*, this verb makes no direct reference to the notion of witnesses. When the speaker names witnesses or guarantors, he generally does so with a prepositional phrase introduced by *per* (e.g. *Lex Bant.* CIL 1^2 582.24; *Lex Salpens.* CIL 2.1963.24). The simple accusative first appears in this function in Cicero's reference to a traditional oath: *Iovem lapidem iurare* (*Fam.* 7.12.2). Subsequently this construction appears in Vergil and later poets (*Aen.* 6.324, 12.197=App. 174).[27]

The earliest literary evidence for the use of *iuro* in oaths comes from Plautus' comedies, which contain four examples of *iurare* followed by *per* and a god's name introducing an oath. Three appear in the *Amphytrio*, which includes this ironic exchange between the slave Sosia and the god Mercury in the guise of Sosia:

> SO. per Iovem iuro med esse neque me falsum dicere.
> ME. at ego per Mercurium iuro tibi Iovem non credere;
> nam iniurato scio plus credet mihi quam iurato tibi
> (435-437).

Notice that the primary objective of these oaths is to affirm the truthfulness of the two speakers, who have just been arguing about which one of them truthfully claims to be Sosia.[28] Sallust, who is the next source chronologically, uses the verb only once and that in the narrative. Bomilcar sends a letter to Nabdalsa reminding him of his oath to aid in the overthrow of Jugurtha: *testari deos per quos iuravisset* (*Jug.* 70.5). The opposition of the two verbs shows a distinction between *testari* and *iurare* (cf. Liv. 1.59.1=App. 144). Although *testari* sometimes means "to swear an oath," Bomilcar here invokes divine witnesses of a previous oath to observe its possible violation. Cicero, in contrast to his frequent use of *testor*, uses the verb *iurare* only once as part of an oath by divine witnesses. It appears in a hypothetical oath, which Cicero describes as typical of political argument:

---

[26] Unlike most *verba testandi*, *iuro* only takes on the force of a *verbum orandi* late and rarely (see TLL 7.2.1.677.12-22). Propertius is the first to use it thus (1.15.35). On derivation from *ius*, "formule religieuse qui a force de loi," see Ernout-Meillet[4] s.v. *ius*.

[27] On the indication of witnesses by means of a simple accusative "*usu sec. gr. novato*," see TLL 7.2.675.46-76.

[28] See also *Amph.* 831-834 and *Mil.* 1420-1422 with its pun on *testes* as testicles as well as witnesses.

> nisi ineptum putarem in tali (*sc.* de philosophia) disputatione id facere quod cum de re publica disceptatur fieri interdum solet, iurarem per Iovem deosque penates me et ardere studio veri reperiendi et ea sentire quae dicerem (*Acad.* 2.65).

Perhaps the reason for the verb's infrequency in Cicero is rhetorical. *Iurare* is more common and less dramatic than *testari*. How much more effective it is to summon the gods as personal witnesses of one's words than merely to swear by them.

Livy uses *iuro* only once in association with an oath in *oratio recta.* Brutus swears to avenge the death of Lucretia in the presence of Collatinus, Lucretius, and Valerius, all of whom swear in like manner after him. Livy uses two introductory verbs: *iuro per sanguinem, vosque, di testes facio me . . . exsecuturum . . .* (1.59.1=App. 144). The reason for the dual clauses is likely the rhetorical effect of intensity created by the use of synonymous phrases.[29] In other passages in Livy, *iurare* occurs in the narrative rather than within the oath itself (2.45.14=App. 146, 2.46.6, 23.9.3, 26.48.12).

In the first of two Vergilian oaths using *iuro,* Aeneas affirms and dramatizes his unwillingness to leave Dido:

> . . . per sidera iuro,
> per superos et si qua fides tellure sub ima est
> invitus, regina, tuo de litore cessi
> (6.458-460=App. 170).

This passage uses the traditional structure of a prepositional phrase. In the second occurrence, however, Vergil employs simple accusative objects to indicate witnesses. Before the Trojan hero's duel with Turnus, Latinus introduces a catalogue of witnesses to his oath with Aeneas:

> Haec eadem, Aenea, terram, mare sidera, iuro
> Latonaeque genus duplex Ianumque bifrontem
> (12.197-198=App. 174).

In addition to *iuro,* Latinus uses two other *verba testandi* in the same oath with no obvious distinction in meaning: *audiat* and *testor* (12.200, 201=App. 174).

---

[29] See Ogilvie (1965) *ad loc.* on the model and creation of this oath. Ogilvie notes that an oath by blood is otherwise unattested in Latin.

obtestor
Liv. 6.14.5, 21.10.3
Aen. 9.260

Unlike *testor* from which it is derived, the compound *obtestor* is primarily a verb of intensive supplication, meaning "to implore."[30] As such, it regularly introduces a jussive noun clause. As the root verb would suggest, *obtestor* includes the notion of witnesses. When, as is frequently the situation, the speaker addresses a request directly to another person (in the accusative), a prepositional phrase with *per* names the witnesses, for example, *per omnes deos te obtestor ut* . . . (Cic. *Att.* 11.2.2).

The earliest occurrence of *obtestor* with reference to gods appears in Plautus, where the verb is accompanied in one passage by a reference to divine witnesses. Aristophontes pleads with Hegio not to destroy the slave Tyndarus:

> AR. per deos atque homines ęgo te optestor, Hegio,
> ne tu istunc hominem perduis (*Capt.* 727-728).

Caesar also uses the verb once with reference to the gods. In his will, Ptolemy asks the Roman people to see that his choice of heirs is observed: *haec uti fierent, per omnis deos perque foedera quae Romae fecisset eodem testamento Ptolomaeus populum Romanum obtestabatur* (*BCiv.* 3.108.5). Cicero uses *obtestor* twice in his speeches and once in his letters with reference to divine witnesses (*Att.* 11.2.2). For example, in the grand peroration of his final speech against Verres, Cicero summons a veritable assembly of gods to hear his plea that the judges vote to condemn:

> nunc te Iuppiter Optime Maxime . . . teque Iuno Regina *etc.* . . . ceteros item deos deasque omnis imploro et obtestor . . . ut . . . quae mea mens in suscipienda causa fuit, fides in agenda, eadem vestra (*sc.* iudices) sit in iudicanda . . . (*Verr.* 5.188; cf. *sim. Dom.* 144-145).

In the one example from Sallust, Sulla invokes Jupiter to witness the perfidy of the Moorish king Bocchus: *Iovem maxumum obtestatus, ut*

---

[30] See TLL s.v. *obtestor* 280.77-84, 281.4, 282.29. See also Ernout-Meillet[4] s.v. *testis*: "*obtestor*: même sens que *attestor*, mais souvent pris dans un sens religieux." Cf. Paul. Fest. 201L.=207M. *obtestatio est cum deus testis in meliorem partem vocatur, detestatio, cum in deteriorem*; cf. Cic. *Balb.* 33.

sceleris atque perfidiae Bocchi testis adesset, . . . (*Jug.* 107.2). Here the idea of invoking a witness is felt to be so weakened in *obtestor* that a more explicit phrase is added (*testis adesset*). *Obtestor* occurs once in a formal prayer: the *carmen devotionis* of Macrobius concludes with the invocation of Tellus and Jupiter as witnesses: *Tellus mater teque Iuppiter obtestor* (*Sat.* 3.9.11).

Livy's history contains two oaths using *obtestor*, which are similar to those of his predecessors in that they introduce requests (6.14.5: *deos hominesque ut*; 21.10.3=App. 156: *per deos ne*). In two other passages, the verb preserves its meaning "to implore" but takes the simple accusative object *fidem* instead of introducing a clause (8.33.23, 2.10.3=App. 145). One passage describes Horatius Cocles' attempt to stay the flight of his fellow soldiers: *deum et hominum fidem obtestans testabatur . . .* (2.10.3=App. 145). Here, as in the Sallustian passage above, the notion of witness in *obtestor* is so weak that Livy couples it with the verb *testor* to introduce a statement.

The verb is not common in poetry.[31] *Obtestor* appears in one Vergilian passage naming divinities as witnesses of a request.[32] Ascanius encourages Nisus and Euryalus to seek Aeneas' return from Pallanteum:

> . . . per magnos, Nise, penatis
> Assaracique larem et canae penetralia Vestae
> obtestor, quaecumque mihi fortuna fidesque est,
> in vestris pono gremiis. revocate parentem,
> reddite conspectum . . . (9.258-262=App. 172).

Other than Vergil, Horace is the only Augustan poet who uses *obtestor*. In this passage, Vulteius, the one-time auctioneer, now bankrupt farmer, begs his patron to return him to his former state:

> quod te per Genium dextramque deosque Penatis
> obsecro et obtestor, vitae me redde priori (*Epist.* 1.7.94-95).

---

[31] Lyne on *Ciris* 273: *Obtestor* "is not generally popular in poetry apart from a rather surprising five examples in Verg.—but all in the second half of *Aen.*"

[32] Serv. *ad Aen.* 9.258: *obtestor adiuro.* Cf. Claud. Don. *ad Aen.* 9.258-260: *obtestari est ob aliquam causam deos testes adhibere cum iureiurando.*

testem facio
Liv. 1.32.6, 1.59.1

The noun *testis* appears in a double accusative construction with the names of divine witnesses after a variety of verbs: *adhibere*; *advocare*; *dare*; *invocare*; *testari*.[33] Only rarely does the verb *facere* introduce such a construction in passages which refer to a deity or deities as witnesses. Ovid uses the phrase in a parodic prayer offered by a deceitful merchant to Mercury, god of merchants and thieves:

> sive ego te feci testem, falsove citavi
> non audituri numina vana Iovis (*Fast.* 5.683-684).

In Livy's history, the phrase appears twice in formulae of oaths. In his oath to avenge the rape of Lucretia, Brutus invokes the gods to witness his promise to pursue Superbus and his family and to allow no king to rule in Rome:

> per hunc . . . sanguinem iuro, vosque, di, testes facio me L. Tarquinium Superbum cum scelerata coniuge et omni liberorum stirpe ferro igni quacumque dehinc vi possim exsecuturum, nec illos nec alium quemquam regnare Romae passurum (1.59.1=App. 144).

In a second passage, the phrase introduces an indirect question, an unusual construction after *verba testandi*. The king Tullus Hostilius instructs envoys from the Alban king Cluilius to carry a message back to their leader:

> "nuntiate," inquit, "regi vestro, regem Romanum deos facere testes, uter prius populus res repetentes legatos aspernatus dimiserit, ut in eum omnes expetant huiusce clades belli" (1.22.7=App. 140).[34]

In two other passages, *testem facere* occurs in the narrative (1.32.6-8= App. 142 introducing *oratio recta*; 34.11.8=App. 162 introducing *a.c.i.*).

---

[33] *Adhibere* (Cic. *Fin.* 2.67); *advocare* (Tac. *Hist.* 4.41); *dare* (e.g. Plaut. *Ps.* 514); *invocare* (Liv. 8.6.1, 35.31.13=App. 164, 39.51.12, 41.25.4); *testari* (Plaut. *Rud.* 1338).

[34] Cf. Dion. Hal. 3.3.5: μαρτύρομαι τοίνυν . . . τοὺς θεούς, οὓς ἐποιησάμεθα τῶν σπονδῶν μάρτυρας . . . .

testor
Liv. 1.32.9
Aen. 3.599, 4.492, 12.201, 12.581

In the context of oaths, *testor* normally means "summon as a witness." The earliest example of *testor* occurs in Plautus' *Rudens*; it is the only occurrence of the verb in Plautus. In a lengthy parody of a formal oath, the pimp Labrax begins his promise to reward the finder of his lost trunk with the invocation of Venus: *Venus . . . testem te testor mihi* (*Rud.* 1338).[35] Terence also contains only one example of the verb *testor*, this time in a colloquial context. The *adulescens* Pamphilus swears that his marital separation is not his fault: *neque mea culpa hoc discidium evenisse, id testor deos* (*Hec.* 476). Not surprisingly, considering their rhetorical function, Cicero's speeches contain many examples of oaths, several introduced by this verb (e.g. *Caecin.* 83, *Manil.* 70). One example will suffice. Near the close of his speech in defense of P. Sulla, Cicero addresses the tutelary gods of the state to guarantee his integrity in arguing the case: *di patrii ac penates, qui huic urbi atque huic rei publicae praesidetis . . . testor integro me animo ac libero P. Sullae causam defendere . . .* (86). There are also two examples in his letters, which illustrate the colloquial usage of oaths in the first century (*Quint.* 1.3.2, *Fam.*10.35.1). Sallust's works contain only one oath introduced by *testor*. The envoys sent by the conspirator C. Manlius to meet the general Q. Marcius Rex begin speaking by invoking gods and men to witness (*deos hominesque testamur*) that they have no intention of harming the country but only of protecting their own rights (*Cat.* 33.1). None of the oaths in these predecessors of Livy and Vergil accompanies requests.

In invocations of divine witnesses, Livy employs the verb *testor* only once within an oath written in *oratio recta*. In the fetial formula for declaring war, the speaker asks the gods to witness the wrongdoing of the other party: *ego vos testor populum illum . . .iniustum esse* (1.32.9-10=App. 143). On all other occasions, this verb appears in the narrative (e.g. *obtestansque deum et hominum fidem testabatur nequiquam deserto praesidio eos fugere. . .* (2.10.3=App. 145).[36] The syntactical context of *testor* in Livy conforms to that of his literary predecessors. In all of these passages, an accusative object of *testor*, either the phrase *deos*

---

[35] See the introduction to this chapter for the entire Plautine oath.

[36] Also 2.10.3, 2.57.4, 3.72.1, 4.53.5, 9.31.10, 22.44.6, 28.8.2.=App. 145, 147, 150, 151, 155, 158, 160. No reason for invoking witnesses is specified at 38.33.7.

hominesque/atque homines or the names of specific gods, indicates the witnesses invoked (9.31.10=App. 155).

In the *Aeneid, testor* occurs in three oaths that are similar to those of predecessors.[37] Dido asks the gods to witness the truth of the statement that she unwillingly resorted to magic:

> testor, cara deos et te, germana, tuumque
> dulce caput, magicas invitam accingier artis
> (4.492-493=App. 168).

Latinus also invokes divine witnesses of his treaty with Aeneas:

> . . . numina testor:
> nulla dies pacem hanc Italis nec foedera rumpet, . . .
> (12.201-202=App. 174).

In a third oath, set in *oratio obliqua*, Aeneas addresses Latinus and seeks divine witnesses of his truthfulness: *testaturque deos iterum se ad proelia cogi* . . . (12.581=App. 175). A fourth oath, however, accompanies a request, Achaemenides' plea for the Trojans to remove him from the island of the Cyclops:

> . . . per sidera testor,
> per superos atque hoc caeli spirabile lumen,
> tollite me, Teucri . . . (3.599-601=App. 167).

This is the earliest example of *testor* accompanying a request and of its usage with a prepositional phrase to refer to witnesses.[38]

II. Formulae Regarding Perjury

An oath sometimes contains a formulaic phrase expressing the trustworthiness of the speaker. For example, Scipio swears *ex mei animi sententia*, "with full knowledge."[39] The speaker may also affirm good

---

[37] In Vergil, as in Livy, a reference to the gods does not always accompany the verb *testari*. In addition to deities, Vergil expands the types of witnesses to include heavenly fires invoked by Sinon (2.154) and Trojan rivers Xanthus and Simois invoked by Jupiter (5.803).

[38] Connington-Nettleship notes that Forcellini does not mention the common meaning "to adjure" which appears here (*ad l.*). *Per deos* does appear with other *verba testandi* in predecessors.

[39] See Cicero's discussion on the requirements for a binding oath (*Off.* 3.108). He singles out the phrase *ex animi sententia* for special comment (see below).

intentions with the formula *sine dolo malo*. The technical, legal formula *sciens dolo malo*, which appears in municipal oaths under Domitian, expresses both of these concerns about a speaker's trustworthiness—that he speak both with knowledge of the meaning of his obligations and with the intention to fulfill those obligations.[40]

None of these technical formulae appear in their entirety in Vergil or other poets.[41] The closest that any Vergilian oath approaches to such a phrase is Latinus' oath to Aeneas:

> . . . nec me vis ulla volentem
> avertet . . . (12.203-204=App. 174).

The idea that Latinus will not "willingly" violate the terms of the agreement may be intended to parallel the technical promise not to deceive "knowingly."

ex animi sententia

Liv. 22.53.10

Ex animi sententia is a technical formula which affirms the speaker's truthfulness by stating understanding and acceptance of the obligations undertaken. Thus Cicero explains:

> quod enim ita iuratum est, ut mens conciperet fieri oportere, id servandum est; quod aliter, id si non fecerit, nullum est periurium . . . Non enim falsum iurare periurare est, sed, quod ex animi tui sententia iuraris, sicut verbis concipitur more nostro, id non facere periurium est (*Off.* 3.107-108).

The phrase also occurs in the A.D. 37 oath of allegiance to Caligula by the people of Aritium and in colloquial oaths (CIL 2.172=ILS 190).[42]

Livy uses the formula once in an oath that invokes the gods. After the disaster at Cannae, P. Scipio swears an oath of fidelity to the Roman state and forces a group of soldiers considering desertion to do likewise.

---

[40] E.g. *Lex Bant.* CIL 1.582.8, 17; *Lex Col. Gen. Iul.* ILS 6087.81; Fest. 247L.= 221M.; see also Kaser 420, 524; Mommsen *StrR* 1.86.

[41] For *si fallo*, see Prop. 2.20.16, 4.7.53, 4.11.27. An Ovidian parody reads *sive deum prudens alium divamve fefelli* (*Fast.* 5.685).

[42] Cf. Cod. Just. 2.58.2.2: *convenit enim et ipsos iureiurando adfici, quia ipsi causam scientes ita ad eam perveniunt . . . ipsi . . . scire possunt causam et ita ad iudicium pervenire eo, quod ex animi sui sententia iurent*. See TLL 2.95.8-19 s.v. *animus* for additional examples of this phrase.

He begins: *ex mei animi sententia . . . ut ego rem publicam populi Romani non deseram neque alium civem Romanum deserere patiar* (22.53.10=App. 159). He then calls on Jupiter Optimus Maximus to punish him and his household with death if he violates his oath. Another oath that does not make reference to the gods also employs this formula (43.15.8).

si sciens fallo
Liv. 22.53.11

Si sciens fallo is a technical formula, which introduces the concluding self-curse of a formal oath. The verb *fallere* points to the primary function of an oath as an affirmation of truthfulness. The use of the participle *sciens* underscores the legal concept that responsibility for an action is dependent upon understanding one's obligations and the legality of one's actions. When no other object is expressed, as is typical, the reader should perhaps understand the object of the verb *fallere* to be *deos* (e.g. Liv. 2.45.13=App. 146, 30.42.21).[43] The complete formula has only survived in a few instances (e.g. Plin. *Pan.* 64.3). According to Paulus' epitome of Festus, the formula is part of the traditional oath by Jupiter Lapis: *Lapidem silicem tenebant iuraturi per Iovem, haec verba dicentes: si sciens fallo, tum me Diespiter salva urbe arceque bonis eiiciat uti ego hunc lapidem* (102L.=115M.).[44] The earliest appearance of the formula in literature occurs in Cicero: (*sc. Aesopus*) *iurare cum coepisset, vox eum defecit in illo loco*: "*si sciens fallo*" (*Fam.* 7.1.2; cf. *Acad.* 2.146: *qui* (*sc. maiores*) *primum iurare "ex animi sententia" quemque voluerunt, deinde ita teneri "si sciens falleret"*). It is part of the concluding self-curse of the oath of Aritium from A.D. 37:

> si s[cie]ns fa[ll]o fefellerove tum me liberosq(ue) meos Iuppiter Optimus Maximus ac Divus Augustus ceteriq(ue) omnes di immortales expertem patria incolumitate fortunisque omnibus faxint (CIL 2.172=ILS 190).

Only once does Livy use the full formula, in the lengthy oath that Scipio forces potential deserters to swear after the crushing defeat at Cannae. He concludes his oath in the traditional manner with a curse of

---

43 See Ogilvie (1965) *ad* 2.45.14 and TLL s.v. *fallo* 6.182.35-39 for additional examples. *Adde* Tac. *Ann.* 1.73. Other objects may occur (e.g. Liv. 29.24.3). None of these passages includes the participle *sciens*.

44 On this oath, see the introduction to this chapter.

himself: *si sciens fallo, tum me Iuppiter optime maxime domum familiam remque meam pessimo leto adficias* (22.53.11=App. 159). In two other oaths in *oratio obliqua* the phrase appears without the participle *sciens* (2.45.14, 21.45.8=App. 146, 157).[45]

sine dolo malo
Liv. 1.24.7

The technical legal formulae *dolo malo* and *sine dolo malo*[46] are related to the previous formula *si sciens fallere* in the sense that guilt is considered to be dependent upon the intent of the agent. The phrase survives in the Republican law of the *Colonia Genetiva Iulia*, which records the oath to be taken by civic scribes:

> ius iurandum adigito per Iovem deosque Penates "sese pecuniam publicam eius colon(iae) concustoditurum rationesque veras habiturum esse, u(ti) q(uod) r(ecte) f(actum) e(sse) v(olet) s(ine) d(olo) m(alo), neque se fraudem per litteras facturum esse sc(ientem) d(olo) m(alo)" (ILS 6087.81).

Livy uses the formula once in an oath. The Roman *pater patratus* declares that he has recited the conditions of the treaty with the Alban people *sine dolo malo* (1.24.7=App. 141; cf. *sine fraude* 1.24.5=App. 141). He then restricts the condition under which Jupiter is asked to punish the Roman people if they violate the treaty: *si prior defexit publico consilio dolo malo, tum . . .* (1.24.8=App. 141). Livy repeats the formula in two other oaths, neither of which specifically name a deity as witness (treaty: 38.11.2, 5; military oath: 43.15.8).[47]

III. Self-Curses

A formal oath regularly concludes with a curse invoked on the speaker in the event he should perjure himself. In Livy, several examples appear in fetial formulae. In accounts of the ratification of treaties, a sacrifice strengthens the curse; the killing of the sacrificial victim serves as a

---

[45] So also the Plautine oath (*Amph.* 933); note Kleinknecht's comment "ein richtiges Gebet" (170).

[46] *Lex Tarent.* CIL 1.590.25=ILS 6086; *Lex Col.Genet. Iul.* ILS 6087 (9x); *Edictum Augusti de aquaeductu Venafrano* CIL 10.4842=ILS 5743.25, ILS 7934. On *dolus malus*, see Kaser 214, 524 and Mommsen *StrR* 1.86.

[47] The latter also contains the phrase *ex tui animi sententia*; see above.

symbol for the killing of the potential perjurer (1.24.8).[48] The formula for the satisfaction of wrongs closes with a curse of exile for the speaker if the demands are unjust: *si . . . tum patriae compotem me numquam siris (sc. Iuppiter) esse* (1.32.8). This recalls the famous oath by Jupiter Lapis, in which the casting away of a stone symbolizes the expulsion of the perjurer from his city (Paul. Fest. 102L.=115M.). Two other oaths invoke divine punishment if the speaker fails to serve his country well in battle; one of these names death as the penalty: *si fallat . . . iratos invocat deos* (2.45.14=App. 146); *si sciens fallo tum me . . . leto adficias* (22.53.10=App. 159).[49] These curses regularly begin with a *si* clause, stating the condition under which the curse is activated. Next follows a clause often beginning with the words *tum* or *ita* and stating the punishment.[50]

ferire
Liv. 1.24.8, 9.5.3

In addition to its general meaning "to strike," *ferire* appears several times in Livy and other authors as a technical term meaning "to solemnize a treaty" by killing a sacrificial pig.[51] The occurrence of the verb in fetial prayers demonstrates the etymological connection between these two usages of *ferire*: *tum tu ille Diespiter populum Romanum sic ferito, ut ego hunc porcum hic hodie feriam; tantoque magis ferito, quanto magis potes pollesque* (Liv. 1.24.8=App. 141).[52] Livy also shows this relationship when he defines a *foedus*: *ubi precatione res transigitur, per quem populum fiat quo minus legibus dictis stetur, ut eum ita Iuppiter*

---

[48] See below under *ferire*. On self-curses, see Hirzel 8. On the symbolic rites of Near Eastern treaty making, see McCarthy 91-95, 150-153.

[49] Cf. 26.25.11 which does not mention the gods. In addition to the passages mentioned in this paragraph, the following also refer to punishment: 2.45.14, 9.5.3, and 21.45.8=App. 146, 154, 157.

[50] *Tum*: Liv. 1.24.8, 1.32.8, 22.53.10=App. 141, 142, 159; *ita*: 9.5.3, 21.10.3= App. 154, 156. When a blessing rather than a curse is invoked, it may also begin with *ita* (*Aen.* 9.208=App. 171). Cf. the English "so help me God."

[51] Cf. *Initiis pacis, foedus cum feritur, porcus occiditur* (Varr. *Rust.* 2.4.9). Poets also occasionally use *ferire* meaning "to ratify a treaty" (Enn. *Ann.* 32; Verg. *Aen.* 10.154). For additional examples in prose and poetry, see TLL 6.516.56-63 s.v. *ferire* and 6.1006.75-79.

[52] Ogilvie (1965) calls this "an archaizing reconstruction" (110). On its rhythm and structure, see Norden (1939) 99, 285. Cf. the similar verb *mactare* Liv. 21.50.8=App. 43 below. N.B. the epithet Feretrius applied to Jupiter who was especially associated with oaths.

feriat quemadmodum a fetialibus porcus feriatur (9.5.3=App. 154). Although there is no documentary evidence for the Latin wording of the fetial oath, the use of *ferire* to identify the solemnization of a treaty may indicate that *ferire* was also a technical term in the oath. Since the legal ceremony is confirmed by a religious sacrifice, it is difficult to distinguish the religious and legal aspects of this terminology.

mactare
Liv. 21.45.8

In one oath Livy substitutes the verb *mactare* for *ferire*. Prior to the battle near Victumulae, Hannibal swears to his troops that he will abide by his promises of rewards: *agnum laeva manu dextra silicem retinens, si falleret, Iovem ceterosque precatur deos ita se mactarent quemadmodum ipse agnum mactasset* . . . (21.45.8=App. 157).[53] There is no evidence to suggest that *mactare* was a technical term used in the *execratio* of an oath.[54] It was, however, a religious term for the act of sacrifice and thus furnished the proper tone for an oath before the gods (e.g. Cato *Rust.* 134.2, 4). Servius refers to it as a *verbum sacrorum* (*ad Aen.* 4.57). The related term *mactus/macte*, which does appear in requests of the gods, means "honored" or "increased" (e.g. Cato *Rust.* 132; CIL 6.32323.98 =Pighi 1141.98).[55]

Asseverations

In addition to the lengthy oaths, Livy's history and the *Aeneid* include briefer asseverative expressions, such as *mediusfidius* and *hercule*. These generally serve as emphatic particles appearing in colloquial or rhetorical contexts. Despite an original religious sense, their usage in Latin literature is usually divorced from any real religious notion. For this reason, I treat them rather summarily.

[53] This oath also differs from others in Livy by the choice of a lamb rather than the usual Roman sacrifice of a pig. On the sheep as the typical Greek sacrificial victim, see Lehr 79. The same Carthaginian practice is evidenced by examples of Neo-Assyrian treaties (Parpola-Watanabe no. 2 ll.10-35 and no. 6 ll. 547-554).

[54] But see Appel, who includes the verb in his study (180-182).

[55] For additional examples of *mactare*, see TLL 8.22.14-45; *mactus/macte*, see TLL 8.23.3953-54 and 69-83.

hercule/hercules/mehercule
Liv. 2.28.4 *et saepius*

This asseveration calls upon the god Hercules as a witness of the speaker's truthfulness. For the most part, only men use *hercule* and its variants, while women prefer *ecastor.*[56] According to Gagnér, *hercule* derives from a simple invocation using the vocative case of the god's name (6). The compound *mehercule,* however, as Festus wrote, is an abbreviated form of the prayer *ita me Hercules iuvet* (112L.=125M.). An earlier form may have contained a vocative with the imperative.[57] Although the asseveration often seems to emphasize certain words or phrases, according to Gagnér, it affirms the entire clause in which it appears (111). The pattern of usage shows two primary contexts—colloquial speech and oratory. Among poets, Plautus and Terence make frequent use of this asseveration in their comedies.[58] It is also frequent in Cicero's works, especially in the speeches and letters. Livy's predecessors in history only rarely use it.[59] In Sallust, all four occurrences appear in speeches.

The asseveration occurs 46 times in Livy.[60] Only on one occasion does Livy employ the expanded form *mehercule* (38.17.18=App. 165). In all but seven passages, the context is a rhetorical or emotional speech. For example, Senators complain of consular inactivity concerning nocturnal gatherings of the plebs: *unum hercule virum—id enim plus esse quam consulem—qualis Ap. Claudius fuerit, momento temporis discussurum illos coetus fuisse* (2.28.4).

---

[56] On the use of this asseveration, see Gagnér 80, 87; Nicholson 99-103.

[57] On early forms, see Gagnér 22, 34.

[58] For locations in Plautus and Terence, see Gagnér 205-210 and 215 n.2. *Adde* Catull. 38.2 *mehercule*; Hor. *Sat.* 2.7.72, *Epist.* 1.15.39.

[59] Caes. *apud* Cic. *Att.* 9.7C.1; Ps. Caes. *BAfr.* 12, 16, *frg.* 18; Sall. *Cat.* 52.35, *Jug.* 85.46, 110.2, *Hist.* 1.77.17; Tac. (20 times).

[60] Four of these occurrences use the form *hercules* (3.68.6, 5.4.10, 28.44.12, 34.31.3).

mediusfidius

Liv. 2.31.9, 5.6.1, 22.59.17, 34.5.13

Like *mehercule, mediusfidius* presupposes some fuller expression such as *ita me Dius Fidius iuvet.*[61] This asseveration calls upon the aid of a divinity known as Dius Fidius, perhaps to be identified with Jupiter in his function as a god of oaths. It does not occur in literature as frequently as either *hercule* or *mehercule*. Neither Plautus nor Terence uses this asseveration. That it belonged to the colloquial speech of the first century is shown by its relative frequency in Cicero, in speeches and more often in his letters (12x, 21x). Two other orators attest its rhetorical usage: Scipio Minor and M. Cato (Scip. min. *orat.* 30; Cato *orat.* 176, *apud* Gell. *NA* 10.14.3). The asseveration occurs only once in Sallust and not at all in Caesar (Sall. *Cat.* 35.2 a letter). Poets avoid it entirely.

Livy uses *mediusfidius* four times, all in the context of speeches. For example, the dictator M'. Valerius addresses the Senate after the failure of a resolution concerning debtors: *optabitis, mediusfidius, propediem, ut mei similes Romana plebes patronos habeat* (2.31.9).

[61] On this asseveration, see Gagnér 20-21. On Dius Fidius, see Wissowa 118, 129-131. Plautus refers to Dius Fidius once: *per Dium Fidium* (*Asin.* 23). See Gagnér 215 on its rarity among early writers of comedy. He cites one occurrence: *mim. anonym.* 4.

VI: Comparisons

In order to illustrate the differences between the religious language of prayers in Livy's history and those in the *Aeneid*, I will examine several passages chosen to exemplify similar types of prayers that occur in both works: oaths at the solemnization of a treaty, vows before an intended action, and petitionary prayers in response to an omen.

Treaty Oaths

In the first book of Livy's history, a Roman fetial priest, the *pater patratus*, summons men and gods to witness the treaty between Rome and the city of Alba Longa, granting victory to the army whose triplet heroes defeat in battle the triplets of the other army:

> Pater patratus . . . multisque id verbis, quae longo effata carmine non operae est referre, peragit. Legibus deinde recitatis, "Audi," inquit, "Iuppiter; audi, pater patrate populi Albani; audi tu, populus Albanus. Ut illa palam prima postrema ex illis tabulis cerave recitata sunt sine dolo malo utique ea hic hodie rectissime intellecta sunt, illis legibus populus Romanus prior non deficiet. Si prior defexit publico consilio dolo malo, tum illo die, Iuppiter, populum Romanum sic ferito ut ego hunc porcum hic hodie feriam; tantoque magis ferito quanto magis potes pollesque" (1.24.6-8=App. 141).

Livy's text contains all the characteristic elements of an actual oath: invocation of witnesses, promise, reference to intent, and self-curse. Livy's *pater patratus* begins with a brief invocation, naming only Jupiter as a divine witness of his oath. This conforms well to the model established by the one extant Latin oath, that of allegiance to Caligula, which invokes Jupiter and the divine Augustus; other deities are mentioned only in a universal invocation. Following the invocation, the priest carefully refers to the terms as they have been previously read. He notes that this statement of terms has been made without intent to deceive. Then creating an escape clause, the *pater patratus* swears that Rome will not be the *first* party to violate the terms. The priest closes with a curse; if the

Romans should violate the terms, he prays that Jupiter will punish them severely and then he strikes the sacrificial pig to ritually symbolize that punishment. The *pater patratus* carefully phrases the curse to include escape clauses in case a violation occurs unintentionally or contrary to public policy.

The language of this treaty is both legal and religious in character, two spheres difficult to distinguish since this religious ritual is part of legal procedure. Its technical terminology, as well as its focus on precision, reflects both spheres, but the references to fraud are especially typical of legal documents (*sine dolo malo, dolo malo*). Interestingly, the verbs *iuro* or *testor* are missing. Instead, Livy uses a verb (*audi*) that, although possibly technical, is not attested in other prose prayers. The repetition of words and phrases (*audi, populus Romanus, populus Albanus, dolo malo, ferito*) and alliteration (*pater patrate populi*; *palam prima postrema*; *potes pollesque*) recall the solemn language of both prayer and law. Since no documentary source for fetial oaths survives, there is no way of knowing how closely those in Livy's history resemble actual oaths in language. They probably do offer a realistic model of their content and tone.[1]

In contrast to the judicio-religious tone of the fetial oath in Livy, the oath that Aeneas swears to solemnize the treaty with the Latins has a distinctly literary tone. Before his combat with Turnus, Aeneas invokes a number of deities to witness his oath:

> Esto nunc Sol testis et haec mihi Terra vocanti,
> quam propter tantos potui perferre labores,
> et pater omnipotens et tu Saturnia coniunx
> (iam melior, iam, diva, precor), tuque inclute Mavors,
> cuncta tuo qui bella, pater, sub numine torques;
> fontisque fluviosque voco, quaeque aetheris alti
> religio et quae caeruleo sunt numina ponto:
> cesserit Ausonio si fors victoria Turno,
> convenit Evandri victos discedere ad urbem,
> cedet Iulus agris, nec post arma ulla rebelles
> Aeneadae referent ferrove haec regna lacessent.
> sin nostrum adnuerit nobis Victoria Martem

[1] Cf. Livy's treatment of the *senatus consultum de Bacchanalibus* and the parallel inscription (39.8-18; CIL 1.196=ILS 18).

(ut potius reor et potius di numine firment),
non ego nec Teucris Italos parere iubebo
nec mihi regna peto: paribus se legibus ambae
invictae gentes aeterna in foedera mittant.
sacra deosque dabo; socer arma Latinus habeto,
imperium sollemne socer; mihi moenia Teucri
constituent urbique dabit Lavinia nomen
(*Aen.* 12.176-194=App. 173).

In turn, Latinus swears:

"Haec eadem, Aenea, terram, mare, sidera, iuro
Latonaeque genus duplex Ianumque bifrontem,
vimque deum infernam et duri sacraria Ditis;
audiat haec genitor qui foedera fulmine sancit.
tango aras, medios ignis et numina testor:
nulla dies pacem hanc Italis nec foedera rumpet,
quo res cumque cadent; nec me vis ulla volentem
avertet, non, si tellurem effundat in undas
diluvio miscens caelumque in Tartara solvat,
ut sceptrum hoc" (dextra sceptrum nam forte gerebat)
"numquam fronde levi fundet virgulta nec umbras,
cum semel in silvis imo de stirpe recisum
matre caret posuitque comas et bracchia ferro,
olim arbos, nunc artificis manus aere decoro
inclusit patribusque dedit gestare Latinis"
(12.197-211=App. 174).

Both Aeneas and Latinus begin with lengthy and ornate invocations. The choice of deities invoked differs considerably from that in the oath of the Aritiensians, which names only Jupiter and the Divine Augustus and mentions other gods only in a general summary phrase (CIL 2.172=ILS 190). Vergil's invocation, according to Lehr, may be modeled on Homeric oaths, for example *Il.* 19.258-265, which invokes Zeus, Gaia, Helios and the Erinyes (58). This invocation is similar to the oath of allegiance to Augustus from Gangra in 3 B.C. (ILS 8781). Witnesses invoked are Jupiter, Gaia, Helios, all gods and goddesses, and Augustus. The invocations of Aeneas and Latinus also bear a resemblance to that which Polybius attributes to the Carthaginian Hannibal in his treaty with Philip of Macedon:

Ἐναντίον Διὸς καὶ Ἥρας καὶ Ἀπόλλωνος, ἐναντίον δαίμονος Καρχηδονίων καὶ Ἡρακλέους καὶ Ἰολάου, ἐναντίον Ἄρεως, Τρίτωνος, Ποσειδῶνος, ἐναντίον θεῶν τῶν συστρατευομένων καὶ Ἡλίου καὶ Σελήνης καὶ Γῆς, ἐναντίον ποταμῶν καὶ λιμένων καὶ ὑδάτων, ἐναντίον πάντων θεῶν ὅσοι κατέχουσι Καρχηδόνα, ἐναντίον θεῶν πάντων ὅσοι Μακεδονίαν καὶ τὴν ἄλλην Ἑλλάδα κατέχουσιν, ἐναντίον θεῶν πάντων τῶν κατὰ στρατείαν, ὅσοι τινὲς ἐφεστήκασιν ἐπὶ τοῦδε τοῦ ὅρκου (Polyb. 7.9.2-3).

Aeneas and Latinus, like Polybius' Carthaginians, swear by the Sun and the Earth and waters, not traditional witnesses of Roman treaties. The relative clauses describing Jupiter in Latinus' oath (*genitor qui foedera fulmine sancit*) and Earth and Mars in Aeneas' oath are poetic expressions of the clauses found in authentic prayers naming the sphere of influence of the deity. Aeneas' parenthetical prayer to Juno (*iam melior, iam, diva, precor*), almost an exclamation, would also be inappropriate in a formal oath. Aeneas specifies the terms of the agreement, but he does not include the customary curse for violation of terms. Although Latinus does not include an actual penalty clause in his oath either, he does refer to such penalties in his identification of Jupiter, *qui foedera fulmine sancit* (*Aen.* 12.200=App. 174). The list of *adynata* in Latinus' oath are poetic in style. The reference to the scepter, although expressed in very poetic language, may recall an actual curse, namely that the speaker be equally lifeless and sterile. Also missing from both oaths is any mention of the symbolic relationship between the slaying of sacrificial animals and a curse on the speakers.

In terms of diction, Vergil has maintained an appropriately formal tone, but has avoided all technical formulae. For verbs of oath-taking, Vergil has once used a technical term *testor* and once substituted the phrase *esto testis*, which recalls Homer's language in the oath preceding the duel of Paris and Menelaus, μάρτυροι ἔστε (*Il.* 3.280). The choice of a future imperative for this latter phrase, as with *habeto* (*Aen.* 12.192), is in keeping with the judicio-religious nature of the oath. In Latinus' oath, Vergil, like Livy, adds the verb *audire*. There are two alliterative phrases (*potui perferre*; *Fontisque Fluviosque*), which contribute to the solemn tone, but since alliteration is a fairly common poetic device in the *Aeneid*, its use here does not distinguish the oath from the language of the poem as a whole.

Vows

The contrast between Livy and Vergil in terms of technical language is not always as pronounced as in the case of fetial oaths. During battle with the Sabines, Romulus vows a temple:

> "Iuppiter, tuis" inquit "iussus avibus hic in Palatio prima urbi fundamenta ieci. Arcem iam scelere emptam Sabini habent; inde huc armati superata media valle tendunt; at tu, pater deum hominumque, hinc saltem arce hostes; deme terrorem Romanis fugamque foedam siste. Hic ego tibi templum Statori Iovi, quod monumentum sit posteris tua praesenti ope servatam urbem esse, voveo" (Liv. 1.12.4-6=App. 111).

Compare this vow, which Ascanius makes before he throws his spear at Remulus:

> Iuppiter omnipotens, audacibus adnue coeptis.
> ipse tibi ad tua templa feram sollemnia dona,
> et statuam ante aras aurata fronte iuuencum
> candentem pariterque caput cum matre ferentem,
> iam cornu petat et pedibus qui spargat harenam
> (*Aen.* 9.625-629=App. 137).

Neither author uses technical formulae in these passages, but Livy does use the technical terms *voveo* and *arce*. Especially noteworthy is Vergil's avoidance of the verb *voveo* in five out of six vows in the *Aeneid* (*excepitur* 10.774=App. 139). He prefers instead the future tense of a verb of offering. In contrast, in the six vows which Livy presents in *oratio recta*, only one contains a future in place of a verb meaning "to vow" (36.2.2=App. 131). The only technical term Vergil employs in Ascanius' prayer is the adjective *aurata*, which appears in vows of the *Fratres Arvales* to describe the sacrificial victim. In the prayers of the Arvals, however, the phrase is *bove aurato* rather than *aurata fronte* (Henzen 110, 122). It is just enough detail to remind the audience of an actual sacrifice.

Both passages contain the essential elements of a vow: invocation, request, and promise. The choice and description of offerings point to characteristic differences between the two authors. Romulus' vow of a temple is in keeping with Livy's interest in describing the historical origins of monuments still to be seen in the Rome of his day. Ascanius' vow, in contrast, has no historical associations at all; rather it offers Vergil the opportunity to elaborate poetically on the appearance and death of the sacrificial victims. In general, Livy's prayer functions as part of the historical narrative and Vergil's calls up a timeless image.

Although the previous vow is typical of those in Vergil, the passage from Livy is not typical for that author. Sixteen of the 23 vows in Livy are brief single-sentence statements; all but one occur in *oratio obliqua.* Typically they contain a technical or pseudo-technical formula of request expressed as a conditional clause. For example, before the war against Perseus in 171 B.C., the consul C. Popilius vows to celebrate games to Jupiter Optimus Maximus: *decrevit senatus C. Popilius consul ludos per dies decem Iovi optimo maximo fieri voveret . . . si res publica decem annos in eodem statu fuisset* (42.28.8=App. 133).

Simple Petitions

Both works include prayers in response to auspicious omens. In Livy, when a raven perches on the helmet of Valerius Corvinus, he responds to the omen by praying: *si divus, si diva esset qui sibi praepetem misisset volens propitius adesset* (7.26.4=App. 29). The language is technical and formulaic (*si divus, si diva*; *volens propitius*). It is also ritually correct. Corvinus has no way of knowing which deity has sent the omen and so he must avoid offending that deity by using the wrong name or assuming the wrong gender.

In the *Aeneid*, after he has seen the flame on Iulus' head, Anchises prays:

> Iuppiter omnipotens, precibus si flecteris ullis,
> aspice nos, hoc tantum, et si pietate meremur,
> da deinde auxilium, pater, atque haec omina firma
> (2.689-691=App. 82).

Several poetic, perhaps Greek, elements appear in this prayer. The two conditional clauses, which argue for the fulfillment of the prayer on the basis of divine compassion and human piety, present a persuasive element, which does not occur in any actual Latin prayer, but is characteristic of Greek and Latin poetry. The verb *aspice* too is not attested in any actual prayer. Even in Latin poetry it seldom occurs, but it recalls the common Greek request ἴδοι. Finally, the requests *da* and *firma* employ the imperative, a form which appears only rarely in actual prayers.

In response to a second omen, Anchises makes only the request *servate domum, servate nepotem* (2.703=App. 83). Although the verb is technical, it does not appear here in the traditional formula, *salvum servare*. In addition, Vergil again employs the imperative.

VII: Conclusions

The examination of petitionary prayers has shown that the language of prayers in Livy and Vergil differs greatly. Differences between the literary prayers of these two authors come into sharper focus when viewed against the background of authentic prayers of Roman cult. Although both authors generally observe the typical structural and conceptual elements of cultic prayer, their treatment of traditional language varies considerably. While the prayers of Livy's history frequently summon up images of official state occasions under Augustus, those of Vergil's *Aeneid* elicit literary images of religious practices that extend beyond narrow boundaries of space and time.

Vergil's *Aeneid*

The most striking characteristic of prayers in the *Aeneid* is the total absence of technical formulae.[1] Admittedly, the restrictions of meter eliminate certain words or combinations of words, but Ennius demonstrates clearly the possibility of adapting religious formulae to dactylic hexameter.[2] Only Aeneas' prayer to the Tiber is phrased in such a way as to suggest that Vergil had in mind a traditional Latin prayer as a model (8.72=App. 136). Although the meaning of that Vergilian prayer is very similar to the one in Ennius (*Ann.* 54), significant differences in diction result in a weakening of the official tone, as is appropriate for the context.

Vergil does use individual words which traditionally appear within technical formulae (e.g. *servare* of the formula *salvum servare*). Because of the separation of the words from traditional formulae, the Augustan audience probably did not associate these words with the prayers of state

---

[1] Lehr 34-35; Highet 120-121.

[2] E.g. *Ann.* 110-111. Although this particular combination of words is unique, both the diction and the arrangement recall actual prayers and the atmosphere of the Roman state with its formulaic liturgy. Cf. *ut ea res mihi collegaeque meo bene et feliciter eveniat* (e.g. Cic. *Mur.* 1; Liv. 40.46.9=App. 73). Ennius has perhaps adopted the adverb *fortunatim* from a variant form of *quod bonum faustum felixque sit* (e.g. *fortunatum*, Cic. *Div.* 1.102). See also Skutsch (1985) 250-252.

cult, but recognized them merely as appropriate for Latin prayers. In some cases, these individual technical terms occur with a change of tone. The synonyms *bonus* and *felix*, which appear in the official formula *quod bonum faustum felixque sit*, retain the meaning "propitious," but each adjective occurs in contexts suggesting an ironic interpretation: Aeneas' request that the disguised Venus be *felix*, (1.330=App. 134) and Dido's invocation of *bona* Iuno (1.734=App. 79). Especially within the context of the ill-fated encounter between Aeneas and Dido, which is arranged by the goddesses Venus and Juno, the use of "propitious" to describe either goddess has ironic possibilities.

Vergil retains the technical meanings and usages of some technical words,[3] but introduces new meanings or usages for other technical vocabulary. *Laetus* and *felix*, which traditionally modify events or objects, describe the gods themselves. The adjectives *dexter*, *placidus*, and *secundus* are not previously attested in prayers. There are no occurrences of *secundus* in any other prayers, in either prose or poetry. Vergil has also removed the technical term *volens* from its usual formula *volens propitius* and changed the object which that adjective usually modifies. Turnus applies it to himself, but in authentic prayers, *volens* always refers to the deity whose help and favor are sought (10.677=App. 104). This usage may hint at the irreligiosity of Turnus, which manifests itself more clearly in the subsequent prayer to his spear (12.95-100).

Vergil's moderation in the use of technical vocabulary is found in both prayer and non-prayer contexts. In his study on epic language in the *Aeneid*, Cordier found that only 3.5% of all verses in the poem contained technical words.[4] Such a low figure reflects a literary movement away from this traditional feature of epic. Specialized vocabulary from varied spheres, including the religious, had lent precision and vitality to the language of Homer and his successors. The *Annales* shows that Ennius accepted that tradition: 11.5% of his verses contain technical language. But during the first century there occurs a decreased usage of technical vocabulary in the epic poems of Cicero (7% technical vocabulary, all from

[3] E.g. *avertere, fas, iuvare, prohibere, sanctus,* and *sinere*. Vergil also uses *fas* with the non-technical meaning "fate" (*Aen.* 6.63=App. 92).

[4] Cordier does not label "technical" many words which I distinguish as technical terms of prayer, as for example, *precor*. This may be due to his unfamiliarity with the specialized language of prayer, and not to the use of a different definition. See Cordier 103-104 for his usage of "le mot technique."

the religious sphere) and a dramatic drop in Lucretius' epic (2%).[5] Cordier's observation concerning the *Aeneid* as a whole, that Vergil uses technical language, but in moderation, is in agreement with the observation of the present study with respect to the language of prayers (95, 103-108, 166).

As with technical language, Vergil's treatment of colloquial prayers shows some interesting variations of formulae. For example, the colloquial *di faciant* becomes the more formal *sic pater ille deum faciat* (10.875-876=App. 105). The brief invocation *pro Iuppiter* is the only colloquial prayer which remains unaltered (4.590=App. 169).

Vergil also avoids an official tone by eschewing the repetitions characteristic of state cult. In contrast to Livy's eight-fold repetition of the prayer *quod bonum faustum felixque sit,* Vergil repeats only one phrase with religious connotations, *pede secundo* (8.302, 10.255=App. 97, 102).[6] Even then, Vergil varies the clauses in which this phrase appears; the difference in vocabulary, as well as in the context and speaker, removes any similarity between these two prayers except for the idea of "propitious."

In several cases, Vergil's chosen vocabulary has no synonyms in actual Latin prayers. Instead, his diction shows a close relationship to the language of Greek prayers. In particular, the verbs requesting divine mercy or pity (*misereri*; *miserescere*) in the prayers of poets (but not of prose sources) recall such Greek petitions as ἐλέαιρε and οἴκτιρε (e.g. Hom. *Od.* 5.450; IG[2] 971; Aesch. *Cho.* 502). The plea for a god to look at a petitioner (*aspice*) is also characteristic of Greek prayers (e.g. ἴδοι Aesch. *Suppl.* 206). *Quicumque* recalls the Greek term ὅτις (e.g. Hom. *Od.* 5.445. The affinity of language to that of Greek prayers is a characteristic Vergil shares with other Roman poets. Sometimes the content of prayers in the *Aeneid* probably prompted the reader or listener to recall Vergil's Greek predecessors. Three prayers in particular suggest models from the Homeric epics.

This apparent use of Greek models may be due in part to subject matter; the characters are, after all, Homeric. Prayers of the Roman state

---

[5] Catullus, however, returns to the frequent use of technical vocabulary in *carmen* 64 (Cordier 108).

[6] The only other repeated phrase in prayers of the *Aeneid* is *crudelem abrumpere vitam* (8.579, 9.497=App. 99, 101).

cult would be anachronistic, although it should be noted that Vergil has included other anachronistic details of Roman cult, such as the opening of the gates of the temple of Janus in times of war (7.607-615). It is also natural that private individuals would not use the language of official Roman cult. Still, the prayers of Cato's agricultural treatise prove that there were similarities between the language of private and public cult. Vergil's dominant concern was to employ a literary language removed from that of daily Latin speech and one which would recall the Greek epics he sought to rival.

Although the primary narrative time is set during the Trojan war and the years immediately following, Vergil carries this story to the present of Augustan Rome and into the future as well by means of visions and prophecies. Similarly, the language of prayers does not bind them firmly to any time period; they exist in a place outside of time, a place of poetry. This timelessness of language is one element which enables the *Aeneid* to continue today as one of the masterpieces of world literature.

Livy's History

The phrasing of prayers in Livy's history represents for the most part the highly formulaic liturgy of Roman state cult. Most religious words in these prayers either belong to technical formulae or are independent technical terms.[7] Although individuals sometimes used the same language in personal prayers, the contexts in which Livy employs this language would recall civil and military experiences to his Augustan audience rather than private ones. This combination of frequent technical formulae and public settings contributes a very formal and official tone to Livy's history.

The repetition of formulaic phrases, both in the narrative and in the prayers themselves, emphasizes this official tone.[8] For example, the prayer *bene vertat*, with negligible variations, appears nine times, *quod*

---

[7] Technical formulae in Livy include *bene ac feliciter eveniat, ex animi sententia, in tutela esse, precor quaesoque, probe factum esto, quod bonum faustum felixque sit, recte factum esto, salvum servare, si divus si diva, si res publica in eodem statu, si sciens fallo, sine dolo malo, veniam peto* and *volens propitiusque*. In addition, Livy uses the following technical terms: *arcere, avertere, dare, devovere, fas, ferire, iurare, iuvare, sanctus, sinere, testor,* and *voveo*. For a survey of literature on Livy's language, see Aili 1122-1147.

[8] See Phillips 193-207 on repetition in official notices.

bonum faustum felixque sit eight times, and *bene ac feliciter eveniat* six times. Some of this repetition reflects Livy's tendency to repeat a phrase he has just used. Thus, *bene vertat* appears three times in book 3, *bene ac feliciter eveniat* three times within four chapters of book 31, and *quod bonum faustum felixque sit* twice each in books 1 and 3. But Livy is careful to alternate the use of the similar prefatory prayers *bene vertat* and *quod bonum faustum felixque sit.*[9] In addition, these closely recurring passages represent only one half of the occurrences of these formulae. For example, *quod bonum faustum felixque sit* also appears in books 8, 10, 24 and 42. When the phrasing and repetition of formulaic prayers throughout Livy is combined with that of annalistic notices, the official tone becomes the dominant one. Together they reminded Livy's contemporary audience that the Roman state, with its political and religious machinery, had been and continued to be a stable and enduring entity.[10]

Livy did not, however, restrict himself to technical formulae. On occasion, he altered this vocabulary to form phrases not attested elsewhere in prayers. Sometimes his variations involve the substitution of adjectives for adverbs or vice versa (e.g. *prosperum evenire*). But Livy did not change the traditional restriction of certain adjectives to either gods or events. Sometimes individual words belonging to technical formulae also appear in non-formulaic phrases. For example, the adjectives *faustus* and *felix*, which usually occur in the official formula *quod bonum faustum felixque sit*, also appear individually and in combination with other adjectives. The adjectives *laetus* and *prosperus*, although technical terms of augury, also occur in prayers in Livy. There is no evidence from prose prayers outside of Livy for such usages. These variations allowed Livy to avoid monotony while still maintaining a ritualistic tone.

Livy's introduction of his history with a pseudo-invocation reminiscent of poets supports the generally accepted view that one important function of the numerous ritual adornments is their contribution to atmosphere. But what is the nature of that atmosphere? Some scholars like Walsh attribute the accounts of liturgy to the antiquarian interests of the Augustan age (175). Although the longest and most striking Livian prayers probably depend on antiquarian works, this influence is relatively

---

[9] On Livy's use of *variatio*, see Phillips 193-207.

[10] On the appeal to "political and religious traditionalism," see McDonald 159 and Phillips 207.

small considering the total number of prayers. Out of 127 prayers examined, only six show the detail indicative of antiquarian interest in ritual: the inaugural prayer of Numa (1.18.9=App. 5), the fetial oaths (1.24.7-8, 1.32.6=App. 141, 142), the *carmen evocationis* (5.21.2-3=App. 21), the *carmen devotionis* (8.9.6-8=App. 115), and the vow of a *ver sacrum* (22.10.2-6=App. 123).[11] Five of these appear in the first decade where there is somewhat more of an antiquarian atmosphere. The fetial oath for the declaration of war had, in fact, something more than an antiquarian interest for the Romans of Livy's day. That ancient ritual was probably used by Augustus in his declaration of war against Cleopatra. A large number of prayers in Livy's history simply and briefly request divine favor in attempted undertakings, usually military and/or civil. They are the sort of prayers which members of Livy's audience had in all probability heard or spoken themselves.

The language of Livy's prayers shows the same preference for the contemporary. With the exception of a few archaisms such as the poetic *sospitare* and *verruncare*, the vast majority of words were probably in use in Livy's day.[12] Livy uses several colloquial formulae which are attested from the late Republic and Augustan Rome.[13] A similar observation may be made with respect to Livy's choice of forms; he again prefers the contemporary to the archaic. Although Cicero and Tacitus attest the continuation of the archaic form *duint* in the colloquial formula *di mentem duint*, Livy chooses the contemporary form in three of five prayers containing the verb *dare*.[14] Similarly, the archaic forms *esto* and *faxitis* only appear in two prayers, both of which have several archaic features.[15] Livy also prefers contemporary orthography and writes *si*

[11] On Livy's limited interest in antiquarian material, see Luce 160-161.

[12] There is evidence for the first century use of the technical *bene ac feliciter eveniat, ex animo sententia, fas, iuro, obtestor, oro, precor quaesoque, probe factum sit, quod bonum faustum felixque sit, recte factum sit, salvum servare, si divus si diva, si res publica in eodem statu, si sciens fallo, sine dolo malo, testor, veniam peto, volens propitiusque,* and *voveo.*

[13] E.g. *bene ac feliciter eveniat, di approbent, di faciant, di meliora, di omen avertant, hercule, mediusfidius,* and *ne istuc Iuppiter Optimus Maximus sirit.*

[14] Cic. *Cat.* 1.22, *Phil.* 10.13; Tac. *Ann.* 4.38; Liv. 3.17.6, 10.24.16, 29.27.4=App. 11, 40, 55. The prayers in Livy containing the archaic form refer to human actions (10.19.17-18, 22.10.3=App. 116, 123).

[15] *Esto* occurs in the vow of a *ver sacrum* (22.10=App. 123) and both appear in Scipio's prayer (29.27.2=App. 55).

divus si diva instead of *sei deivus sei deiva,* the spelling found on inscriptions. This preference for contemporary language in prayers is consistent with Livy's diction throughout his history.[16]

Although Livy attributes all of the prayers we have examined to speakers who lived centuries earlier and although the prayers probably resembled those of centuries past, Livy's audience must have recognized the similarities to contemporary ritual. Some formulae, such as the prefatory prayer *quod bonum faustum felixque sit,* had probably continued in ritual use with little change for hundreds of years. The Augustan reader, coming across such a prayer in the account of an assembly which took place seven hundred years earlier, would not consider that prayer antiquarian, but rather appropriate and even obvious words for the occasion. They themselves had heard speakers pray *quod bonum faustum felixque sit* before addressing an assembly. They had heard magistrates in state rituals ask that the gods be *volentes propitiique* and that the action to be undertaken turn out *bene ac feliciter* for the Roman people. The language was familiar. The atmosphere thus created by the prayers was not primarily of antiquity, but of the living present.

Despite these differences in language, both Livy's history and the *Aeneid* illustrate a similar view of Roman history. The creation and preservation of the Roman state represent the shared efforts of gods and men. Individual men and the state as a corporate body are, in part, responsible for the success both of that state and of themselves, to the degree that they manifest piety. Pious Aeneas emerges victorious over the impious Turnus and Mezentius. The prosperity of the Roman state also, according to Camillus, reflects its piety towards the gods: "observe indeed then the events of the past years, both favorable and unfavorable, you will discover that all things have prospered for those who followed the gods, but for those who spurned them, all things were unfavorable" (Liv. 5.51.4-5). Prayers represent one way of manifesting the desired piety. The pious and thus successful actors of the *Aeneid* and of Livy's history are those who are in frequent communication with the gods by means of prayer. Aeneas offers more prayers than any other individual in the *Aeneid.* The impious Mezentius and Turnus both offer prayers to their weapons (10.773-776=App. 139, 12.95-100). The majority of prayers in Livy's history are spoken by persons representing the Roman state, such as

[16] On Livy's use of contemporary diction, see Gries 6-7.

consuls, generals, and senators. But prayer is not just a technique of characterization, it is a necessary feature of a world in which the *pax deum* is essential for success. Prayer is the means by which Romans request omens and aid. It accompanies sacrifices and divination. Neither epic history nor epic verse would be realistic without them.

Appendices

Appendix 1: Petitionary Prayers
Livy

1. 1.*praef.*13: Livy to his readers:[1]

> . . . cum **bonis** potius ominibus votisque et precationibus deorum dearumque, si, ut poetis, nobis quoque mos esset, libentius inciperemus, ut orsis tantum operis successus **prosperos** darent.

2. 1.16.3: Soldiers at apotheosis of Romulus:

> . . . pacem precibus exposcunt, uti **volens propitius** suam semper **sospitet** progeniem.

3. 1.16.6: Proculus Iulius to the deified Romulus:

> . . . petens precibus ut contra intueri **fas** esset, . . .

4. 1.17.10: *Interrex* to electoral *contio*:

> Tum interrex contione advocata "**quod bonum faustum felixque sit**" inquit "Quirites regem create . . ."

5. 1.18.9: Augur inaugurating Numa Pompilius:

> . . . precatus ita est: "Iuppiter pater, si est **fas** hunc Numam Pompilium cuius ego caput teneo regem Romae esse, uti tu signa nobis certa adclarassis inter eos fines quod feci."

6. 1.28.1: Tullus Hostilius orders the Alban and Roman camps united:

> **Quod bene vertat**, castra Albanos Romanis castris iungere iubet . . .

7. 1.28.7: Tullus Hostilius to the Albans:

> . . . **quod bonum faustum felixque sit** populo Romano ac mihi vobisque, Albani, populum omnem Albanum Romam traducere in animo est . . .

---

[1] Spelling and punctuation follow the Oxford Classical Texts for Livy and Vergil with the exception that "v" has been used for consonantal "u." Weissenborn-Muller furnishes the text for prayers from Livy, books 36-42. Bold print indicates that the words or formulae are discussed in the text.

8. 2.6.8: Arruns Tarquinius before attacking L. Iunius Brutus:

> . . . Di regum ultores **adeste**.

9. 2.10.11: Horatius Cocles as he leaps off the Pons Sublicius:

> Tum Cocles "Tiberine pater," inquit, "te sancte **precor**, haec arma et hunc militem **propitio** flumine accipias."

10. 2.49.7: Well-wishers to the Fabii as they march against the Veientes:

> Praetereuntibus Capitolium arcemque et alia templa, quidquid deorum oculis, quidquid animo occurrit, precantur ut illud agmen **faustum** atque **felix** mittant, sospites brevi in patriam ad parentes restituant.

11. 3.17.6: P. Valerius at close of speech to Quirites:

> Romule Pater, tu mentem tuam, qua quondam arcem ab his iisdem Sabinis auro captam recepisti, da stirpi tuae; iube hanc ingredi viam, quam tu dux, quam tuus ingressus exercitus est. Primus en ego consul, quantum mortalis deum possum, te ac tua vestigia sequar.

12. 3.26.9: Q. Cincinnatus to messengers:

> . . . rogatus ut, quod bene verteret ipsi reique publicae, togatus mandata senatus audiret . . .

13. 3.34.2: *Decemviri* to a *contio* after assembling the 12 Tables:

> . . . populum ad contionem advocaverunt et, **quod bonum faustum felixque** rei publicae ipsis liberisque eorum **esset**, ire et legere leges propositas iussere: . . .

14. 3.35.8: *Decemvir* Appius Claudius promises to convene the *comitia*:

> Ille enimvero, **quod bene vertat**, habiturum se comitia professus, impedimentum pro occasione arripuit; . . .

15. 3.54.8: *Legati* to the plebs who had seceded:

> . . . **quod bonum faustum felixque sit** vobis reique publicae, redite in patriam ad penates coniuges liberosque vestros.

16. 3.62.5: Consul M. Horatius to his troops regarding battle with the Sabines:

Postquam ingenti alacritate clamor est sublatus, **quod bene vertat** gesturum se illis morem posteroque die in aciem deducturum adfirmat.

17. 4.2.8: M. Genucius and C. Curtius consuls concerning the plebeian election of consuls:

Ne id Iuppiter optimus maximus **sineret** regiae maiestatis imperium eo recidere; . . .

18. 4.13.14: Cincinnatus before being named dictator:

. . . precatus tandem deos immortales Cincinnatus ne senectus sua in tam trepidis rebus damno dedecorive rei publicae esset, dictator a consule dicitur.

19. 4.46.4: Q. Servilius concerning strife between two tribunes during war with the Labicani:

. . . precatus ab dis immortalibus ne discordia tribunorum damnosior rei publicae esset quam ad Veios fuisset, . . .

20. 5.18.11-12: Matrons during Veientine war:

concursumque inmuros est iet matronarum . . . obsecrationes in templis factae, precibusque ab dis petitum ut exitium ab urbis tectis templisque ac moenibus Romanis **arcerent** Veiosque eum **averterent** terrorem, si sacra renovata rite, si procrata prodigia essent.

21. 5.21.2-3: *Evocatio* of Veientine Juno by dictator M. Furius Camillus:

"Tuo ductu," inquit, "Pythice Apollo, tuoque numine instinctus pergo ad delendam urbem Veios, tibique hinc decimam partem praedae **voveo**. Te simul, Iuno regina, quae nunc Veios **colis**, **precor**, ut nos victores in nostram tuamque mox futuram urbem sequare, ubi te dignum amplitudine tua templum accipiat."

22. 5.21.15: Camillus after seeing Veientine booty:

. . . precatus esse ut si cui deorum hominumque nimia sua fortuna populique Romani videretur, ut eam invidiam lenire quam minimo suo privato incommodo publicoque populi Romani liceret.

23. 5.32.9: Camillus as he departs into exile:

. . . precatus ab dis immortalibus si innoxio sibi ea iniuria fieret, primo quoque tempore desiderium sui civitati ingratae facerent.

24. 6.20.9: M. Manlius in his own defense:

. . . Iovem deosque alios devocasse ad auxilium fortunarum suarum precatusque esse ut, quam mentem sibi Capitolinam arcem protegenti ad salutem populi Romani dedissent, eam populo Romano in suo discrimine darent, . . .

25. 6.23.11: Camillus yielding to L. Furius' opinion in the Volscian War:

. . . id a dis immortalibus precari ne qui casus suum consilium laudabile efficiat.

26. 6.26.6: Tusculan dictator to Senate:

Haec mens nostra est—di immortales faciant—tam **felix** quam pia.

27. 6.29.2: Cincinnatus at close of speech to Atratinus, *magister equitum,* before battle with Praenestean army:

Adeste, di testes foederis, et expetite poenas debitas simul vobis violatis nobisque per vestrum numen deceptis.

28. 7.20.3: Envoys of the Caerites to the Roman people:

. . . deos rogaverunt, quorum sacra bello Gallico accepta rite procurassent, ut Romanos florentes ea sui misericordia caperet quae se rebus adfectis quondam populi Romani cepisset; . . .

29. 7.26.4: M. Valerius Corvinus in response to an auspicious omen before combat with a Gaul:

Precatus deinde, **si divus, si diva** esset qui sibi praepetem misisset, **volens propitius adesset**.

30. 7.39.13: Mutinous soldiers choose Cincinnatus as their general:

Nomine audito extemplo agnovere virum et, **quod bene verteret**, acciri iusserunt.

31. 7.40.4-5: M. Valerius Corvinus to mutinous army:

"Deos" inquit "immortales, milites, vestros (publicos) meosque ab urbe proficiscens ita **adoravi veniamque** supplex poposci ut mihi de vobis concordiae partae gloriam non victoriam darent . . ."

32. 8.5.6: L. Annius Setinus, Latin envoy, at close of speech to Senate:

". . . **quod** utrisque **bene vertat**, sit haec sane patria potior et Romani omnes vocemur."

33. 8.25.10: Charilaus, envoy of Palaepolis, to P. Philo consul:

. . . **quod bonum faustum felix** [sic] Palaepolitanis populoque Romano **esset**, tradere se ait moenia statuisse.

34. 9.8.8-10: Sp. Postumius at end of speech to the Senate concerning the Caudine Peace:

Vos, di immortales, **precor quaesoque**, si vobis non fuit cordi Sp. Postumium T. Veturium consules cum Samnitibus prospere bellum gerere, at vos satis habeatis vidisse nos sub iugum missos, vidisse sponsione infami obligatos, videre nudos vinctosque hostibus deditos, omnem iram hostium nostris capitibus excipientes; novos consules legionesque Romanas ita cum Samnite gerere bellum **velitis**, ut omnia ante nos consules bella gesta sunt.

35. 9.9.6: Spurius Postumius resumes his speech to the Senate:

. . . quid tandem, si spopondissemus urbem hanc relicturum populum Romanum? si incensurum? si magistratus, si senatum, si leges non habiturum? si sub regibus futurum? **Di meliora**, inquis.

36. 9.14.8: Consuls respond to Samnite declaration that they will not fight:

Accipere se omen consules aiunt et eam precari mentem hostibus ut ne vallum quidem defendant.

37. 10.8.12: Decius Mus at close of speech to *contio*:

Ego hanc legem, **quod bonum faustum felixque sit** vobis ac rei publicae, uti rogas, iubendam censeo.

38. 10.13.12: Q. Fabius Maximus upon election as consul:

". . . **dei approbent**" inquit, "quod agitis acturique estis, Quirites."

39. 10.18.14: L. Volumnius, consul, in response to Ap. Claudius' rejection of his aid:

Bene, Hercules, **verteret**, dicere Volumnius; . . .

40. 10.24.16: P. Decius, consul, in speech to *contio*:

Iovem optimum maximum deosque immortales se precari, ut ita sortem aequam sibi cum collega dent si eandem virtutem felicitatemque in bello administrando daturi sint.

41. 10.35.14: Marcus Atilius, consul, to soldiers in response to a Samnite attempt to fence in the Roman camp:

Facerent—**quod di bene verterent**—quod se dignum quisque ducerent; . . .

42. 21.17.4: Propitiatory *supplicatio* for the war against Carthage:

. . . adorati di, ut **bene ac feliciter eveniret** quod bellum populus Romanus iussisset.

43. 21.50.8: Hiero to Ti. Sempronius:

. . . precatusque **prosperum** ac **felicem** in Siciliam transitum, . . .

44. 23.13.4: Hanno on his mention of adverse fortune:

Quod si, id **quod di omen avertant**, nunc quoque fortuna aliquid variaverit, . . .

45. 24.16.9: Ti. Gracchus to volunteer slaves:

. . . **quod bonum faustum felixque** rei publicae ipsisque esset, omnes eos liberos esse iubere. Ad quam vocem cum [clamor ingenti alacritate sublatus esset ac nunc complexi inter se gratulantesque, nunc manus ad caelum tollentes] bona omnia populo Romano Gracchoque ipsi precarentur, . .

46. 24.21.10: Syracusans remove weapons from temple of Jupiter:

Precantes Iovem ut **volens propitius** praebeat sacra arma pro patria, pro deum delubris, pro libertate sese armantibus.

47. 24.38.8: L. Pinarius in address to troops in Sicily:

Vos, Ceres mater ac Proserpina, **precor**, ceteri superi infernique di, qui hanc urbem, hos sacratos lacus lucosque colitis, ut ita nobis **volentes propitii adsitis**, si vitandae, non inferendae fraudis causa hoc consilii capimus.

48. 26.9.8: Roman women at news of Hannibal's entry into Fregellae:

Matronae . . . circa deum delubra discurrunt . . . orantesque ut urbem Romanam e manibus hostium eriperent matresque Romanos et libros parvos inviolatos servarent.

49. 26.41.16: Hannibal retreating from Rome:

. . . Hannibal, . . . nihil iam maius precatur deos quam ut incolumi cedere atque abire ex hostium terra liceat.

50. 27.45.8-9: Romans concerning C. Claudius Nero's anticipated encounter with Hasdrubal:

Deos omnes deasque precabantur ut illis **faustum** iter, **felix** pugna, matura ex hostibus victoria esset, damnarenturque ipsi votorum quae pro iis suscepissent ut, quemadmodum nunc solliciti prosequerentur eos, ita paucos post dies laeti ovantibus victoria obviam irent.

51. 28.28.11: P. Scipio in speech to mutinous army at Sucro:

Ne istuc Iuppiter optimus maximus **sirit**, urbem auspicato dis auctoribus in aeternum conditam huic fragili et mortali corpori aequalem esse.

52. 28.41.13: Quintus Fabius Maximus to senate:

Quid? si—**quod** omnes **di omen avertant** . . . —victor Hannibal ire ad urbem perget, . . .

53. 29.14.13: Romans welcome the Idaean Mother:

. . . precantibus ut **volens propitia**que urbem Romanam iniret, . . .

54. 29.22.5: Praetor M. Pomponius and *legati* after seeing P. Scipio's battle preparations:

. . . iuberentque **quod di bene verterent** traicere . . .

55. 29.27.2-4: Scipio setting out from Lilybaeum to invade Africa:

". . . divi divaeque" inquit "qui maria terrasque **colitis**, vos **precor quaesoque** uti quae in meo imperio gesta sunt geruntur postque gerentur, ea mihi populo plebique Romanae sociis nominique Latino qui populi Romani quique meam sectam imperium auspiciumque terra mari amnibusque sequuntur **bene verruncent**, eaque vos omnia **bene iuvetis**, bonis auctibus **auxitis**; salvos incolumesque victis perduellibus victores spoliis decoratos praeda onustos triumphantesque mecum domos reduces **sistatis**; inimicorum hostiumque ulciscendorum copiam faxitis; quaeque populus Carthaginiensis in civitatem nostram facere molitus est, ea ut mihi populoque Romano in civitatem Carthaginiensium exempla edendi facultatem detis."

56. 29.27.9: Scipio as he approaches Africa:

. . . Scipio . . . precatus deos uti **bono** rei publicae suoque Africam viderit.

57. 31.5.3-4: Decree of the Senate in 200:

. . . senatusque decrevit uti consules maioribus hostiis rem divinam facerent quibus diis ipsis viderentur cum precatione ea, "Quod senatus populusque Romanus de re publica deque ineundo novo bello in animo haberet, ea res uti populo Romano sociisque ac nomini Latino **bene ac feliciter eveniret**; . . ."

58. 31.7.15: P. Sulpicius Galba at close of speech to a *contio*:

Dii immortales, qui mihi sacrificantique precantique ut hoc bellum mihi, senatui vobisque, sociis ac nomini Latino, classibus exercitibusque nostris **bene ac feliciter eveniret**, laeta omnia prosperaque portendere, . . .

59. 31.8.2: Propitiatory *supplicatio* before war with Philip:

Supplicatio inde a consulibus in triduum ex senatus consulto indicta est, obsecratique circa omnia pulvinaria di ut quod bellum cum Philippo populus iussisset, id **bene ac feliciter eveniret**; . . .

60. 34.24.2: Aristaenus, Achaean praetor, in response to Alexander of Aetolia:

"**Ne** istuc" inquit "Iuppiter optimus maximus **sirit** Iunoque regina, **cuius in tutela** Argi sunt, ut illa civitas inter tyrannum Lacedaemonium et latrones Aetolos praemium sit posita in eo discrimine."

61. 34.34.2: T. Quinctius Flamininus to allies:

"**Bene vertat**" inquit, "obsideamus Lacedaemonem, quando ita placet."

62. 35.18.7: Alexander the Acarnanian at council concerning war:

Meminisse etiam se quotiens in bello precari omnes deos solitus sit ut Antiochum sibi darent adiutorem; . . .

63. 36.1.2: Senate directs consuls to sacrifice and pray before war with Antiochus:

. . . iusserunt precarique, quod senatus de novo bello in animo haberet, ut ea res senatui populoque Romano **bene atque feliciter eveniret.**

64. 36.7.21: Hannibal at end of speech to King Antiochus:

Dii approbent eam sententiam, quae tibi optima visa fuerit.

65. 37.36.6: P. Scipio Africanus to envoy from Antiochus:

. . . aliis, deos **precor,** ne umquam fortuna egeat mea; . . .

66. 37.47.5: Senate decrees propitiatory *supplicatio* after encampment of Scipio's army in Asia in 190 B.C.:

. . . supplicatio decreta est, . . . ut ea res **prospera** et **laeta eveniret.**

67. 38.51.10: P. Scipio Africanus to *contio* in response to charges of malfeasance:

. . . orate deos ut mei similies principes habeatis, . . .

68. 39.10.2: Hispala to Aebutius in response to his intended initiation into the Bacchic rites:

Id ubi mulier audivit, perturbata "**dii meliora!**" inquit; . . .

69. 39.10.5: Hispala to Aebutius:

. . . pacem veniamque precata deorum dearumque, si coacta caritate eius silenda enuntiasset, ancillam se ait dominae comitem id sacrarium intrasse, . . .

70. 39.11.7: Aebutia to consul Postumius regarding her nephew Aebutius:

. . . eiectus a matre, quod probus adulescens—dii **propitii** essent—obscenis, ut fama esset, sacris initiari nollet.

71. 40.4.12: Poris in attempt to save Theoxena and his children from Philip:

. . . deos, ut ferrent opem, orabat.

72. 40.9.5: Perseus to Demetrius (sons of King Philip):

. . . precati tantum deos, ut a me coeptum scelus in me finem habeat . . .

73. 40.46.8-9: Q. Caecilius Metellus in speech to censors-elect:

. . . quod in omnibus fere precationibus nuncupabitis verbis "ut ea res mihi collegaeque meo **bene et feliciter eveniat**, . . ."

74. 40.11.5: Perseus to Demetrius and his father:

Unus ante me pater est, et ut diu sit deos rogo.

75. 42.13.12: King Eumenes in speech to the Senate:

. . . quid ultra facere possum, quam uti ei deos deasque **precor**, ut vos et vestrae rei publicae et nobis sociis atque amicis, qui ex vobis pendemus consulatis?

76. 42.28.7: Senate directs consuls-elect to pray concerning the war against King Perseus:

Consulibus designatis imperavit senatus ut . . . precarentur, ut, quod bellum populus Romanus in animo haberet gerere, ut id **prosperum eveniret**.

77. 42.30.10: Senate directs consuls to present to *comitia centuriata* a resolution of war against Perseus:

[Patres, **quod bonum faustum**] **felixque** populo Romano esset, centuriatis comitiis primo quoque die ferre ad populum consules iusserunt, . . .

Vergil's *Aeneid*

78. 1.603-605: Aeneas to Dido:

di tibi, si qua pios respectant numina, si quid
usquam iustitiae est et mens sibi conscia recti,
praemia digna ferant . . .

79. 1.731-735: Dido at banquet for Trojans:

. . . Iuppiter, hospitibus nam te **dare** iura loquuntur,
hunc laetum Tyriisque diem Troiaque profectis
esse **velis**, nostrosque huius meminisse minores.
adsit laetitiae Bacchus dator et **bona** Iuno;
et vos o coetum, Tyrii, celebrate faventes.

80. 2.190-191: Sinon to Trojans:

. . . tum magnum exitium (**quod di** prius **omen** in ipsum
convertant!) Priami imperio Phrygibusque futurum . . .

81. 2.535-539: Priam curses Pyrrhus:

". . . at tibi pro scelere," exclamat, "pro talibus ausis
di, si qua est caelo pietas quae talia curet,
persolvant grates dignas et praemia reddant
debita, qui nati coram me cernere letum
fecisti et patrios foedasti funere vultus."

82. 2.689-691: Anchises in response to portent of flame on Iulus' head:

. . . Iuppiter omnipotens, precibus si flecteris ullis,
aspice nos, hoc tantum, et si pietate meremur,
da deinde auxilium, pater, atque haec omina firma.

83. 2.701-703: Anchises after confirmation of omen:

Iam iam nulla mora est; sequor et qua ducitis adsum,
di patrii; **servate** domum, **servate** nepotem.
vestrum hoc augurium, vestroque in numine Troia est.

84. 3.34-36: Aeneas after portent of blood dripping from roots of myrtle over the body of Polydorus:

Multa movens animo Nymphas **venerabar** agrestis
Gradivumque patrem, Geticis qui **praesidet** arvis,
rite **secundarent** visus omenque levarent.

85. 3.85-89: Aeneas consulting Apollo of Thymbra:

. . . da propriam, Thymbraee, domum; da moenia fessis
et genus et mansuram urbem; **serva** altera Troiae
Pergama, reliquias Danaum atque immitis Achilli.
quem sequimur? quove ire iubes? ubi ponere sedes?
da, pater, augurium atque animis inlabere nostris.

86. 3.261-266: Response to prophecy of the harpy Celaeno:

. . . sed votis precibusque iubent exposecere pacem,
sive deae seu sint dirae obscenaeque volucres.
et pater Anchises passis de litore palmis
numina magna vocat meritosque indicit honores:
"Di, **prohibete** minas; di, talem **avertite** casum
et **placidi servate** pios . . ."

87. 3.528-529: At the sight of Italy, Anchises responds:

. . . di maris et terrae tempestatumque potentes,
ferte viam vento facilem et spirate **secundi**.

88. 3.619-620: Achaemenides describing the Cyclops:

. . . ipse arduus, altaque pulsat
sidera (di talem terris **avertite** pestem!) . . .

89. 4.576-579: Aeneas in response to the dream vision of Mercury:

. . . sequimur te, **sancte** deorum,
quisquis es, imperioque iterum paremus ovantes.
adsis o **placidus**que **iuves** et sidera caelo
dextra feras . . .

90. 4.607-621: Dido's curse of Aeneas:

Sol, qui terrarum flammis opera omnia lustras,
tuque harum interpres curarum et conscia Iuno,
nocturnisque Hecate triviis ululata per urbes
et Dirae ultrices et di morientis Elissae,
accipite haec, meritumque malis advertite numen
et nostras **audite** preces. si tangere portus
infandum caput ac terris adnare necesse est,
et sic fata Iovis poscunt, hic terminus haeret,
at bello audacis populi vexatus et armis
finibus extorris, complexu avulsus Iuli

auxilium imploret videatque indigna suorum
funera; nec, cum se sub leges pacis iniquae
tradiderit, regno aut optata luce fruatur,
sed cadat ante diem mediaque inhumatus harena.
haec **precor**, hanc vocem extremam cum sanguine fundo.

91. 5.687-692: Aeneas in response to the burning of the Trojan ships:

. . . Iuppiter omnipotens, si nondum exosus ad unum
Troianos, si quid pietas antiqua labores
respicit humanos, da flammam evadere classi
nunc, pater, et tenuis Teucrum res eripe leto.
vel tu, quod superest, infesto fulmine morti,
si mereor, demitte tuaque hic obrue dextra.

92. 6.56-65: Aeneas at the Cumean oracle of Apollo:

. . . Phoebe, gravis Troiae semper **miserate** labores,
Dardana qui Paridis derexti tela manusque
corpus in Aeacidae, magnas obeuntia terras
tot maria intravi duce te penitusque repostas
Massylum gentis praetentaque Syrtibus arva:
iam tandem Italiae fugientis prendimus oras.
hac Troiana tenus fuerit fortuna secuta;
vos quoque Pergameae iam **fas** est parcere genti,
dique deaeque omnes, quibus obstitit Ilium et ingens
gloria Dardaniae . . .

93. 6.196-197: Aeneas in search for a golden branch prays to Venus:

. . . tuque, o, dubiis ne defice rebus,
diva parens . . .

94. 6.264-267: Narrator before describing the underworld:

Di, quibus imperium est animarum, umbraeque silentes
et Chaos et Phlegethon, loca nocte tacentia late,
sit mihi **fas** audita loqui, sit numine vestro
pandere res alta terra et caligine mersas.

95. 6.529-530: Deiphobus to Aeneas curses the Greeks for his own murder:

. . . di, talia Grais
instaurate, pio si poenas ore reposco.

96. 7.259-260: Latinus recalls the oracle of Faunus:

. . . di nostra incepta secundent
auguriumque suum! . . .

97. 8.293-302: Hymn of the Salian priests:

. . . tu nubigenas, invicte, bimembris
Hylaeumque Pholumque manu, tu Cresia mactas
prodigia et vastum Nemeae sub rupe leonem.
te Stygii tremuere lacus, te ianitor Orci
ossa super recubans antro semesa cruento;
nec te ullae facies, non terruit ipse Typhoeus
arduus arma tenens; non te rationis egentem
Lernaeus turba capitum circumstetit anguis.
salve, vera Iovis proles, decus addite divis,
et nos et tua **dexter adi** pede sacra **secundo**.

98. 8.483-484: Evander curses Mezentius:

Quid memorem infandas caedes, quid facta tyranni
effera? di capiti ipsius generique reservent!

99. 8.572-583: Evander as his son Pallas departs for war:

At vos, o superi, et divum tu maxime rector
Iuppiter, Arcadii, **quaeso, miserescite** regis
et patrias **audite** preces: si numina vestra
incolumem Pallanta mihi, si fata reservant,
si visurus eum vivo et venturus in unum,
vitam oro, patior quemvis durare laborem.
sin aliquem infandum casum, Fortuna, minaris,
nunc, nunc o liceat crudelem abrumpere vitam,
dum curae ambiguae, dum spes incerta futuri,
dum te, care puer, mea sola et sera voluptas,
complexu teneo, gravior neu nuntius auris
vulneret . . .

100. 9.404-409: Nisus attempting to rescue Euryalus from the Rutulians:

. . . tu, dea, tu praesens nostro succurre labori,
astrorum decus et nemorum Latonia custos.
si qua tuis umquam pro me pater Hyrtacus aris
dona tulit, si qua ipse meis venatibus auxi

suspendive tholo aut sacra ad fastigia fixi,
hunc **sine** me turbare globum et rege tela per auras.

101. 9.495-497: Mother at the death of her son Euryalus:

. . . aut tu, magne pater divum, **miserere**, tuoque
invisum hoc detrude caput sub Tartara telo,
quando aliter nequeo crudelem abrumpere vitam.

102. 10.252-255: Aeneas in response to portent of ships turned into nymphs:

. . . alma parens Idaea deum, cui Dindyma cordi
turrigeraeque urbes biiugique ad frena leones,
tu mihi nunc pugnae princeps, tu rite propinques
augurium Phrygibusque **adsis** pede, diva, **secundo**.

103. 10.460-463: Pallas before assailing Turnus:

. . . per patris hospitium et mensas, quas advena adisti,
te precor, Alcide, coeptis ingentibus **adsis**.
cernat semineci sibi me rapere arma cruenta
victoremque ferant morientia lumina Turni.

104. 10.676-679: Turnus after miraculous rescue from battle onto a ship:

. . . vos o potius **miserescite**, venti;
in rupes, in saxa (**volens** vos Turnus **adoro**)
ferte ratem saevisque vadis immittite syrtis,
quo nec me Rutuli nec conscia fama sequatur.

105. 10.875-876: Aeneas in response to challenge of Mezentius:

. . . sic pater ille deum faciat, sic altus Apollo!
incipias conferre manum!

106. 11.483-485: Latin women fearing Turnus:

. . . armipotens, praeses belli, Tritonia virgo,
frange manu telum Phrygii praedonis, et ipsum
pronum sterne solo portisque effunde sub altis.

107. 11.557-560: Metabus consecrates his daughter Camilla to Diana:

. . . alma, tibi hanc, nemorum cultrix, Latonia virgo,
ipse pater famulam voveo; tua prima per auras
tela tenens supplex hostem fugit. accipe, testor,
diva tuam, quae nunc dubiis committitur auris.

108. 11.785-793: Arruns prays before assailing Camilla:

. . . summe deum, sancti custos Soractis Apollo,
quem primi colimus, cui pineus ardor acervo
pascitur, et medium freti pietate per ignem
cultores multa premimus vestigia pruna,
da, pater, hoc nostris aboleri dedecus armis,
omnipotens. non exuvias pulsaeve tropaeum
virginis aut spolia ulla peto, mihi cetera laudem
facta ferent; haec dira meo dum vulnere pestis
pulsa cadat, patrias remeabo inglorius urbes.

109. 12.646-649: Turnus forcing battle:

. . . vos o mihi, Manes,
este **boni**, quoniam superis aversa voluntas.
sancta ad vos anima atque istius inscia culpae
descendam magnorum haud umquam indignus avorum.

110. 12.777-779: Turnus pursued by Aeneas:

". . . Faune, **precor, miserere**" inquit "tuque optima ferrum
Terra tene, colui vestros si semper honores,
quos contra Aeneadae bello fecere profanos."

Appendix 2: Vows

Livy

111. 1.12.4-6: Romulus vows temple during battle with Sabines:

> . . . Iuppiter, tuis . . . iussus avibus hic in Palatio prima urbi fundamenta ieci. Arcem iam scelere emptam Sabini habent; inde huc armati superata media valle tendunt; at tu, pater deum hominumque, hinc saltem **arce** hostes; deme terrorem Romanis fugamque foedam siste. Hic ego tibi templum Statori Iovi, quod monumentum sit posteris tua praesenti ope servatam urbem esse, **voveo.**

112. 5.19.6: Camillus vows games and a temple before his campaign against Veii:

> . . . dictator . . . ludos magnos ex senatus consulto **vovit** Veiis captis se facturum aedemque Matutae Matris refectam dedicaturum.

113. 5.21.2-3: Vow of tithe to Apollo and *evocatio* of Veientine Juno by Camillus before battle:

> Tuo ductu, . . . Pythice Apollo, tuoque numine instinctus pergo ad delendam urbem Veios, tibique hinc decimam partem praedae **voveo.** Te simul, Iuno regina, quae nunc Veios **colis, precor,** ut nos victores in nostram tuamque mox futuram urbem sequare, ubi te dignum amplitudine tua templum accipiat.

114. 7.11.4: Dictator Q. Servilius Ahala vows *ludi magni* before campaign against Gauls:

> ex auctoritate patrum, **si prospere** id bellum **evenisset,** ludos magnos **vovit.**

115. 8.9.6-8: *Devotio* of Decius Mus:

> . . . Iane, Iuppiter, Mars pater, Quirine, Bellona, Lares, Divi Novensiles, Di Indigetes, Divi, quorum est potestas nostrorum hostiumque, Dique Manes, vos **precor veneror, veniam peto feroque** uti populo Romano Quiritium vim

victoriam **prosperetis**, hostesque populi Romani Quiritium terrore formidine morteque adficiatis. Sicut verbis nuncupavi, ita pro re publica <populi Romani> Quiritium, exercitu, legionibus, auxiliis populi Romani Quiritium, legiones auxiliaque hostium mecum Deis Manibus Tellurique **devoveo.**

116. 10.19.18: Appius vows temple during battle with Samnites:

. . . Bellona, **si** hodie nobis victoriam duis, ast ego tibi templum **voveo.**

117. 10.28.15-17: *Devotio* of Decius f. P. Decius:

Devotus inde eadem precatione eodemque habitu quo pater P. Decius ad Veserim bello Latino se iusserat **devoveri,** cum secundum sollemnes precationes adiecisset prae se agere sese formidinem ac fugam caedemque ac cruorem, caelestium inferorum iras, contacturum funebribus diris signa tela arma hostium, locumque eundem suae pestis ac Gallorum ac Samnitium fore,—haec exsecratus in se hostesque, . . .

118. 10.36.11: Consul M. Atilius vows temple during battle against the Samnites:

. . . voce clara, ita ut exaudiretur, templum Iovi Statori **vovet, si** constitisset a fuga Romana acies redintegratoque proelio cecidisset vicissetque legiones Samnitium . . .

119. 10.42.7: Consul L. Papirius vows wine during battle with Samnites:

. . . in ipso discrimine quo templa deis immortalibus voveri mos erat **voverat** Iovi Victori, si legiones hostium fudisset, pocillum mulsi priusquam temetum biberet sese facturum.

120. 21.21.9: Hannibal makes vows to Hercules at Gades before war season:

. . . Herculi vota exsolvit novisque se obligat votis, **si cetera prospera evenissent.**

121. 21.62.10: C. Atilius Serranus praetor makes vow to expiate prodigies in accordance the Sibylline Books:

. . . C. Atilius Serranus praetor vota suscipere iussus, **si** in decem annos **res publica eodem stetisset statu.**

122. 22.9.10: Decemvirs report need for *ver sacrum* after Trasimene in accordance with Sibylline Books:

> . . . rettulerunt . . . ver sacrum **vovendum, si** bellatum prospere esset **resque publica in eodem** quo ante bellum fuisset **statu** permansisset.

123. 22.10.2-6: Question of a *ver sacrum* presented to people:

> Velitis iubeatisne haec sic fieri? **Si** res publica populi Romani Quiritium ad quinquennium proximum, sicut velim **<vov>eamque, salva servata** erit hisce duellis, quod duellum populo Romano cum Carthaginiensi est quaeque duella cum Gallis sunt qui cis Alpes sunt, tum donum duit populus Romanus Quiritium quod ver attulerit ex suillo ovillo caprino bovillo grege quaeque profana erunt Iovi fieri, ex qua die senatus populusque iusserit. Qui faciet, quando volet quaque lege volet facito; quo modo faxit **probe factum esto**. Si id moritur quod fieri oportebit, profanum esto, neque scelus esto. Si quis rumpet occidetve insciens, ne fraus esto. Si quis clepsit, ne populo scelus esto neve cui cleptum erit. Si atro die faxit insciens, **probe factum esto**. Si nocte sive luce, si servus sive liber faxit, **probe factum esto**. Si antidea senatus populusque iusserit fieri ac faxitur, eo populus solutus liber esto.

124. 29.36.8: Consul P. Sempronius vows temple before battle with Hannibal near Croton:

> Consul principio pugnae aedem Fortunae Primigeniae **vovit,** si eo die hostes fudisset; composque eius voti fuit. Fusi ac fugati Poeni . . .

125. 30.2.8: Vow of Dictator T. Manlius Torquatus for *ludi magni*:

> . . . consulibus imperatum ut, priusquam ab urbe proficiscerentur, ludos magnos facerent quos T. Manlius Torquatus dictator in quintum annum **vovisset, si eodem statu res publica staret**.

126. 30.27.11: Dictator T. Manlius Torquatus vows *ludi magni*:

> . . . *sc.* (decretum est) ludos quos M. Claudio Marcello, T. Quinctio consulibus T. Manlius dictator quasque hostias maiores **voverat, si** per quinquennium **res publica eodem**

statu fuisset, ut eos ludos consules ad bellum proficiscerentur facerent.

127. 31.21.12: L. Furius Pupurio praetor vows shrine during battle with Gauls at Cremona:

(*sc.* L. Furius Pupurio) aedemque Diiovi **vovit,** si eo die hostes fudisset.

128. 32.6.7: Consul Villius vows temple during battle with Philip in Chaonia near Aous river:

Valerius Antias . . . tradit . . . aedem etiam Iovi in eo proelio **votam, si res prospere gesta esset.**

129. 32.30.10: Consul C. Cornelius Cethegus vows shrine before battle with Insubres near Mincius river:

Consul principio pugnae **vovit** aedem Sospitae Iunono, **si** eo die hostes fusi fugatique fuissent; a militibus clamor sublatus compotem voti consulem se facturos . . .

130. 35.1.9: Praetor P. Cornelius Scipio vows games during battle with Lusitani near Ilipa:

In hoc discrimine ludos Iovi, si fudisset cecidissetque hostes, praetor **vovit**.

131. 36.2.2-5: Consul M'. Acilius vows *ludi magni* before war with Antiochus:

. . . senatus consultum factum est, quod populus Romanus eo tempore duellum iussisset esse cum rege Antiocho . . . utique M'. Acilius consul ludos magnos Iovi **voveret** et dona ad omnia pulvinaria. Id **votum** in haec verba praeeunte P. Licinio pontifice maximo consul **nuncupavit: "Si** duellum, quod cum rege Antiocho sumi populus iussit, id ex sententia senatus populique Romani confectum erit, tum tibi, Iuppiter, populus Romanus ludos magnos dies decem continuos faciet, donaque ad omnia pulvinaria dabuntur de pecunia, quantam senatus decreverit. Quisquis magistratus eos ludos, quando ubique faxit, hi ludi **recte facti** donaque data **recte sunto.**"

132. 41.21.11: People vow *feriae* and *supplicatio* in accordance with the Sibylline Books:

Ex decreto eorum (*sc.* senatus) diem unum supplicatio fuit, et Q. Marcio Philippo verba praeeunte populus in foro votum concepit, si morbus pestilentiaque ex agro Romano emota esset, biduum ferias ac supplicationem se habiturum.

133. 42.28.8: Consul C. Popilius vows games before war against Perseus in 171 B.C.:

. . . decrevit senatus C. Popilius consul ludos per dies decem Iovi optimo maximo fieri **voveret . . . si res publica** decem annos **in eodem statu fuisset.**

Vergil's *Aeneid*

134. 1.327-334: Aeneas vows sacrifices to unrecognized Venus in Africa:

o quam te memorem, virgo? namque haud tibi vultus
mortalis, nec vox hominem sonat; o, dea certe
(an Phoebi soror? an Nympharum sanguinis una?),
sis felix nostrumque leves, **quaecumque**, laborem
et quo sub caelo tandem, quibus orbis in oris
iactemur doceas: ignari hominumque locorumque
erramus vento huc vastis et fluctibus acti.
multa tibi ante aras nostra cadet hostia dextra.

135. 5.234-238: Cloanthus vows sacrifices before ship race:

fudissetque preces divosque in vota vocasset:
"di, quibus imperium est pelagi, quorum aequora curro,
vobis **laetus** ego hoc candentem in litore taurum
constituam ante aras **voti reus**, extaque salsos
proiciam in fluctus et vina liquentia fundam."

136. 8.71-78: Aeneas vows sacrifices in response to dream vision of the god of the river Tiber:

Nymphae, Laurentes Nymphae, genus amnibus unde est,
tuque, o Thybri tuo genitor cum flumine sancto,
accipite Aenean et tandem **arcete** periclis.
quo te cumque lacus miserantem incommoda nostra
fonte tenent, quocumque solo pulcherrimus exis,
semper honore meo, semper celebrabere donis

corniger Hesperidum fluvius regnator aquarum.
adsis o tantum et propius tua numina firmes.

137. 9.624-629: Ascanius vows sacrifices before combat:

. . . ante Iovem supplex per vota precatus:
"Jupiter omnipotens, audacibus **adnue** coeptis.
ipse tibi ad tua templa feram sollemnia dona,
et statuam ante aras aurata fronte iuuencum
candentem pariterque caput cum matre ferentem,
iam cornu petat et pedibus qui spargat harenam."

138. 10.420-423: Pallas vows spoils before assailing Halaesus:

da nunc, Thybri pater, ferro, quod missile libro,
fortunam atque viam duri per pectus Halaesi.
haec arma exuviasque viri tua quercus habebit.

139. 10.773-776: Mezentius vows trophy before assailing Aeneas:

dextra mihi deus et telum, quod missile libro,
nunc **adsint! voveo** praedonis corpore raptis
indutum spoliis ipsum te, Lause, tropaeum
Aeneae.

Appendix 3: Oaths

Livy

140. 1.22.7: Tullus Hostilius to Alban ambassadors:

. . ."Nuntiate," inquit, "regi vestro regem Romanum **deos facere testes**, uter prius populus res repetentes legatos aspernatus dimiserit, ut in eum omnes expetant huiusce clades belli."

141. 1.24.7-8: Sp. Fusis, pater patratus, solemnizes treaty with the Albans:

. . . "**Audi**," inquit, "Iuppiter; **audi**, pater patrate populi Albani; **audi** tu, populus Albanus. Ut illa palam prima postrema ex illis tabulis cerave recitata sunt **sine dolo malo** utique ea hic hodie rectissime intellecta sunt, illis legibus populus Romanus prior non deficiet. Si prior defexit publico consilio **dolo malo**, tum tu illo die, Iuppiter, populum Romanum sic **ferito** ut ego hunc porcum hic hodie **feriam**; tantoque magis **ferito** quanto magis potes pollesque."

142. 1.32.6-8: Fetial formula for demanding satisfaction of wrongs:

. . . "**Audi**, Iuppiter," inquit; "**audite**, fines"—cuiuscumque gentis sunt nominat;—"**audiat fas**. Ego sum publicus nuntius populi Romani; iuste pieque legatus venio, verbisque meis fides sit." Peragit deinde postulata. Inde **Iovem testem facit**: "Si ego iniuste impieque illos homines illasque res dedier mihi exposco, tum patriae compotem me numquam **siris** esse."

143. 1.32.9-10: Fetial formula for the declaration of war:

. . . "**Audi**, Iuppiter, et tu, Iane Quirine, dique omnes caelestes, vosque terrestres vosque inferni, **audite**; ego vos **testor** populum illum"—quicumque est, nominat—"iniustum esse neque ius persolvere; sed de istis rebus in patria maiores natu consulemus, quo pacto ius nostrum adipiscamur."

144. 1.59.1: Brutus over the body of Lucretia:

> . . . "Per hunc" inquit "castissimum ante regiam iniuriam sanguinem **iuro, vosque, di, testes facio** me L. Tarquinium Superbum cum scelerata coniuge et omni liberorum stirpe ferro igni quacumque dehinc vi possim exsecuturum, nec illos nec alium quemquam regnare Romae passurum."

145. 2.10.3: Horatius Cocles to soldiers battling Etruscans:

> . . . **obtestansque deum et hominum fidem testabatur** nequiquam deserto praesidio eos fugere; . . .

146. 2.45.14: Centurion M. Flavoleius before battle against Etruscans:

> "Victor," inquit, . . . "revertar ex acie." **Si fallat**, Iovem patrem Gradivumque Martem aliosque iratos invocat deos. Idem deinceps omnis exercitus in se quisque iurat.

147. 2.57.4: Appius Claudius addresses the Senate:

> Appius contra **testari deos atque homines** rem publicam prodi per metum ac deseri; . . .

148. 3.25.8: Envoy complaining of Aequian treaty violations:

> . . . "Et haec" inquit "sacrata quercus et **quidquid** deorum est **audiant** foedus a vobis ruptum, nostrisque et nunc querellis **adsint** et mox armis, cum deorum hominumque simul violata iura exsequemur."

149. 3.67.7: Consul T. Quinctius Capitolinus at a *contio* of the people:

> **Pro deum fidem**, quid vobis vultis?

150. 3.72.1: Consuls in response to speech of P. Scaptius:

> . . . **deos hominesque testantes** flagitium ingens fieri, patrum primores arcessunt.

151. 4.53.5: Consuls M'. Aemilius and C. Valerius Potitus concerning the obstruction of a levy by the plebeian tribune M. Menenius:

> . . . consulibus **deos hominesque testantibus** quidquid ab hostibus cladis ignominiaeque aut iam acceptum esset aut immineret culpam penes Menenium fore . . .

152. 6.29.2: Dictator Quinctius Cincinnatus at close of speech to A. Sempronius Atratinus, *magister equitum*, before battle with Praenestean army:

Adeste, di testes foederis, et expetite poenas debitas simul vobis violatis nobisque per vestrum numen deceptis.

153. 8.5.8: Consul Titus Manlius in response to Annius, Latin envoy:

. . ."**Audi**, Iuppiter, haec scelera" inquit; "**audite**, Ius Fasque. . . Haecine foedera Tullus, Romanus rex, cum Albanis, patribus vestris, Latini, haec L. Tarquinius vobiscum postea fecit?"

154. 9.5.3: Livy addresses the question of the Caudine Peace:

Quid enim aut sponsoribus in foedere opus esset aut obsidibus, ubi precatione res transigitur, per quem populum fiat quo minus legibus dictis stetur, ut eum ita Iuppiter **feriat** quemadmodum a fetialibus porcus **feriatur**?

155. 9.31.10: Consul C. Iunius Bubulcus exhorts troops going into battle against Samnites:

. . . Iovem Martemque atque alios **testatur deos** se nullam suam gloriam inde sed praedam militi quaerentem in eum locum devenisse neque in se aliud quam nimiam ditandi ex hoste militis curam reprehendi posse; ab eo se dedecore nullam rem aliam quam virtutem militum vindicaturam.

156. 21.10.2-3: Hanno addresses Carthaginian Senate:

Hanno . . . **per deos** foederum arbitros ac testes **obtestans**, ne Romanum cum Saguntino suscitarent bellum . . .

157. 21.45.8: Hannibal addresses his troops:

Eaque ut rata scirent fore, agnum laeva manu dextra silicem retinens, **si falleret**, Iovem ceterosque precatur deos ita se **mactarent** quemadmodum ipse agnum **mactasset**, et secundum precationem caput pecudis saxo elisit.

158. 22.44.5-6: Consul C. Varro to Consul L. Paulus:

Cum . . . **testareturque deos hominesque** hic nullam penes se culpam esse, quod Hannibal iam velut usu cepisset Italiam.

159. 22.53.10: P. Scipio after Cannae:

. . . "**Ex mei animi sententia,**" inquit, "ut ego rem publicam populi Romani non deseram, neque alium civem Romanum deserere patiar; **si sciens fallo,** tum me, Iuppiter optime maxime, domum, familiam remque meam pessimo leto adficias."

160. 28.8.2: Philip addresses council:

. . . **testatus deos hominesque** se nullo loco nec tempore defuisse quin ubi hostium arma concrepuissent eo quanta maxima posset celeritate tenderet . . .

161. 34.5.13: L. Valerius in support of the Lex Oppia:

Superbas, **me dius fidius,** aures habemus si, cum domini servorum non fastidiant preces, nos rogari ab honestis feminis indignamur!

162. 34.11.8: Ambassadors from the Ilergetes to the consul:

. . . **deos hominesque** se **testes facere** invitos et coactos se, ne eadem quae Saguntini passi sint patiantur, defecturos et cum ceteris potius Hispanis quam solos perituros esse.

163. 35.19.3: Hannibal to Antiochus:

. . . Pater Hamilcar . . . altaribus admotum (*sc.* me) iureiurando adegit numquam amicum fore populi Romani.

164. 35.31.13: T. Quinctius before the Magnetes:

Quinctius quidem adeo exarsit ira ut manus ad caelum tendens **deos testes** ingrati ac perfidi Magnetum **invocaret.**

165. 38.17.18: Consul Cn. Manlius to soldiers:

Vobis **mehercule,** Martiis viris, cavenda ac fugienda quam primum amoenitas est Asiae; . . .

166. 44.38.10: Consul L. Aemilius Paulus to council:

Quis **pro deum fidem** ita comparatus . . . (*sc.* non) vicerit.

Vergil's *Aeneid*

167. 3.598-601: Achaemenides to the Trojans on island of the Cyclops:

> . . . mox sese ad litora praeceps
> cum fletu precibusque tulit; "per sidera **testor**,
> **per superos** atque hoc caeli spirabile lumen,
> tollite me, Teucri" . . . .

168. 4.492-493: Dido to her sister Anna:

> **testor**, cara **deos** et te, germana, tuumque
> dulce captut, magicas invitam accingier artis.

169. 4.590-591: Dido seeing the empty port:

> . . . "**pro Iuppiter**! ibit
> hic," ait "et nostris inluserit advena regnis?"

170. 6.458-460: Aeneas to Dido in the underworld:

> funeris heu tibi causa fui? per sidera **iuro**,
> **per superos** et si qua fides tellure sub ima est,
> invitus, regina, tuo de litore cessi.

171. 9.207-209: Nisus in response to Euryalus' protestation of his own bravery:

> Nisus ad haec: "equidem de te nil tale verebar,
> nec fas, non: **ita** me referat tibi magnus ovantem
> Iuppiter aut **quicumque** oculis haec aspicit aequis."

172. 9.257-261: Ascanius promises to reward Nisus for summoning Aeneas:

> "immo ego vos, cui sola salus genitore reducto,"
> excipit Ascanius "**per magnos**, Nise, **penatis**
> **Assaracique larem** et canae penetralia Vestae
> **obtestor**, quaecumque mihi fortuna fidesque est,
> in vestris pono gremiis. revocate parentem, . . ."

173. 12.176-194: Aeneas' oath before his combat with Turnus:

> Tum pius Aeneas stricto sic ense **precatur**:
> "**esto** nunc Sol **testis** et haec mihi Terra vocanti,
> quam propter tantos potui perferre labores,
> et pater omnipotens et tu Saturnia coniunx

(iam **melior**, iam, diva, **precor**), tuque inclute Mavors,
cuncta tuo qui bella, pater, sub numine torques;
fontisque fluviosque voco, quaeque aetheris alti
religio et quae caeruleo sunt numina ponto:
cesserit Ausonio si fors victoria Turno,
convenit Evandri victos discedere ad urbem,
cedet Iulus agris, nec post arma ulla rebelles
Aeneadae referent ferrove haec regna lacessent.
sin nostrum adnuerit nobis victoria Martem
(ut potius reor et potius di numine firment),
non ego nec Teucris Italos parere iubebo
nec mihi regna peto: paribus se legibus ambae
invictae gentes aeterna in foedera mittant.
sacra deosque dabo; socer arma Latinus habeto,
imperium sollemne socer; mihi moenia Teucri
constituent urbique dabit Lavinia nomen."

174. 12.197-211: Latinus' oath before the battle of Aeneas and Turnus:

"haec eadem, Aenea, terram, mare, sidera, **iuro**
Latonaeque genus duplex Ianumque bifrontem,
vimque deum infernam et duri sacraria Ditis;
audiat haec genitor qui foedera fulmine sancit.
tango aras, medios ignis et numina **testor**:
nulla dies pacem hanc Italis nec foedera rumpet,
quo res cumque cadent; nec me vis ulla volentem
avertet, non, si tellurem effundat in undas
diluvio miscens caelumque in Tartara solvat,
ut sceptrum hoc" (dextra sceptrum nam forte gerebat)
"numquam fronde levi fundet virgulta nec umbras,
cum semel in silvis imo de stirpe recisum
matre caret posuitque comas et bracchia ferro,
olim arbos, nunc artificis manus aere decoro
inclusit patribusque dedit gestare Latinis."

175. 12.581-582: Aeneas to Latinus:

testaturque deos iterum se ad proelia cogi,
bis iam Italos hostis, haec altera foedera rumpi.

Appendix 4: Distribution of Prayers

A: Vergil's *Aeneid*

Book	Petitionary Prayer	Vows	Oaths	Total Prayers	Total Lines
1	2	1	0	3	756
2	4	0	0	4	804
3	5	0	1	6	718
4	2	0	2	4	705
5	1	1	0	2	871
6	4	0	1	5	901
7	1	0	0	1	817
8	3	1	0	4	731
9	2	1	2	5	818
10	4	2	0	6	908
11	3	0	0	3	915
12	2	0	3	5	952
Totals	33	6	9	48	9896

B: Livy's History

Book	Petitionary Prayers	Vows	Oaths	Total Prayers
1	7	1	5	13
2	3	0	3	6
3	6	0	3	9
4	3	0	1	4
5	4	2	0	6
6	4	0	1	5
7	4	1	0	5
8	2	1	1	4
9	3	0	2	5
10	5	4	0	9
Total	41	9	16	66
21	2	2	2	6
22	0	2	2	4
23	1	0	0	1
24	3	0	0	3
25	0	0	0	0
26	2	0	0	2
27	1	0	0	1
28	2	0	1	3
29	4	1	0	5
30	0	2	0	2
Total	15	7	5	27

(continued on next page)

Livy's History (continued)

Book	Petitionary Prayers	Vows	Oaths	Total prayers*
31	3	1	0	4
32	0	2	0	2
33	0	0	0	0
34	2	0	2	4
35	1	1	2	4
36	2	1	0	3
37	2	0	0	2
38	1	0	1	2
39	3	0	0	3
40	4	0	0	4
Total	18	5	5	28
41	0	1	0	1
42	3	2	0	5
43 *frg.*	0	0	0	0
44 *frg.*	0	0	1	1
45 *frg.*	0	0	0	0
Total	3	3	1	7
Totals for all decades	77	24	27	128

* The following approximate statistics are taken from Packard v:

Books	*Words*	*Books*	*Words*
1-5	88,400	31-35	62,600
6-10	70,800	36-40	74,300
21-25	76,000	41-45	56,200
26-30	76,700	total:	505,000

Select Bibliography

The following works are cited in the text by the author's name. The date of publication is added when multiple works by the same author are cited.

Aili, H. "Livy's Language. A Critical Survey of Research." *ANRW* II 30.2 (1982) 1122-1147.

Alvar, Jaime. "La formula de la evocatio y su presencia en contextos desacralizadores." *AEA* 57 (1984) 143-148.

"Matériaux pour l'étude de la formule *sive deus, sive dea*." *Numen* 32.2 (1985) 236-273.

Aly, Wolf. *Livius und Ennius*. Leipzig 1936.

Anderson, William S. ed. with commentary. *Ovid's Metamorphoses 6-10*. Norman, Oklahoma 1972.

Appel, Georg. "De Romanorum Precationibus." *Religionsgeschichtliche Versuche und Vorarbeiten*. Vol. 7. Gissa 1909.

Audollent, A. *Defixionum tabellae*. Diss. Paris 1904.

Ausfeld, Karl. *De Graecorum Precationibus Quaestiones*. Leipzig 1903.

Bailey, Cyril. *Religion in Vergil*. Oxford 1935.

Barré, Michael L. *The God-list in the Treaty between Hannibal and Philip V of Macedon: A Study in Light of the Ancient Near Eastern Treaty Tradition*. Baltimore, Maryland 1983.

Basanoff, Vsevolod. *Evocatio; étude d'un rituel militaire romain*. Bibliothèque de l'École des Hautes Études Sciences Religieuses. Vol. 61. Paris 1947.

Bayet, Jean. *Croyances et rites dans la Rome antique*. Paris 1971.

Benveniste, Émile. *Le vocabulaire des institutions indo-européennes*. Paris 1969.

Berretoni, P. "Sopra una formula piaculare nel de agricultura di Catone." *Studi e Sagg. linguist.* 7 (1967).

Birch, R. A. "Livy 22.50 and the Quotations of Ennius." *Latomus* 39 (1980) 88-94.

Block, Elizabeth. *The Effects of Divine Manifestation on the Reader's Perspective in Vergil's Aeneid.* Monographs in Classical Studies. New York 1981.

Bömer, Franz, ed. with comm. *Die Fasten.* Heidelberg 1957.

Boyancé, Pierre. *La religion de Vergil. Mythes et Religiones.* No. 48. Paris 1963.

Bremer, J. M. "Greek Hymns." In *Faith, Hope, and Worship: Aspects of Religious Mentality in the Ancient World,* ed. by Versnel, 193-215. Leiden 1981.

Briscoe, John. *A Commentary on Livy, Books XXXIV-XXXVII.* Oxford 1981.

Cameron, A. "The Date and Identity of Macrobius." *JRS* (1966) 25-38.

Campbell, J. B. *The Emperor and the Roman Army, 31 B.C.-A.D. 235.* Oxford 1984.

Cardauns, Burkhart. *M. Ter. Varro: Antiquitates Rerum Divinarum.* Vol. 1. Wiesbaden 1976.

Champlin, E. "Serenus Sammonicus." *HSCP* 85 (1981) 189-212.

Cipriano, Palmira. *Fas e nefas.* Biblioteca di Ricerche Linguistiche e Filologische. No. 7. Rome 1978.

Conington, John and Henry Nettleship, edd. *The Works of Virgil with a Commentary.* 2 vols. London[5] 1898; reprint 1963.

Cordier, André. *Études sur le vocabulaire épique dans l'Énéide.* Collection d'Études Latines. No. 16. Paris 1939.

Cumont, Franz. "Un serment de fidélité á l'empereur Auguste." *REG* 14 (1901) 26-45.

Daly, Lloyd W. "*Vota publica pro salute alicuius.*" *TAPA* 81 (1950) 164-168.

De-Marchi, Attilio. *Il Culto Privato di Roma Antica.* Milan 1896; reprint 1975.

Devoto, G. *Tabulae Iguvinae.* Rome 1937, 1940[2].

Diehl, E. "Das Saeculum, seine Riten und Gebete." *RhM* 83 (1934). Part 1, 255-272. Part 2, 348-372.

Duckett, Eleanor S. *Studies in Ennius. Monographs of Bryn Mawr College*. Vol. 16. Diss. Bryn Mawr 1904.

Dumézil, Georges. *Idées Romaines*. Bibliothèque des sciences humaines. Paris[2] 1979.

Archaic Roman Religion. Trans. P. Krapp. Chicago 1970.

Dunbabin, R. L. "Verses in Livy." *CR* (1911) 104-106.

Engelbrecht, A. "Zwei alte Gebetsformeln bei Macrobius." *WS* 24 (1902) 478-484.

Erkell, Harry. *Augustus, felicitas, fortuna: lateinische Wortstudien*. Göteborg 1952.

Ernout, Alfred. *Philologica*. 2 vols. *Études et commentaires*. Nos. 1 and 26. Paris 1946.

Ernout, Alfred and Meillet, A. edd. *Dictionnaire étymologique de la langue latine*. Paris[4] 1959.

Forcellini, A., Furlanetto, I., Corradini, F., and Perin, I. *Lexicon Totius Latinitatis*. Patavia 1940.

Fowler, W. *The Religious Experience of the Roman People*. London 1911.

Fraenkel, Eduard. *Plautinisches im Plautus*. Berlin 1922.

Agamemnon. Oxford 1950.

Horace. Oxford 1957.

Friedlander, Ludwig. ed. with comm. *Petronii Cena Trimalchionis*. Leipzig[2] 1906; reprint Amsterdam 1960.

Frier, Bruce Woodward. *Libri Annales Pontificum Maximorum: The Origins of the Annalistic Tradition*. MAAR Vol. 27. Rome 1979.

Fugier, Huguette. *Recherches sur l'expression du sacré*. Publication de la Faculté des lettres de l'Université de Strasbourg. Fascicle 146. Paris 1963.

Fugner, *Lexicon Livianum*. Vol. 1. Leipzig 1897.

Gagnér, A., *De hercle mehercle ceterisque id genus particulis priscae poesis latinae scaenicae*. Greifswald 1920.

Goold, G. "Servius and the Helen Episode." *HSCP* 74 (1970) 102-117.

Gries, Konrad. *Constancy in Livy's Latinity*. Ann Arbor 1949.

Guittard, Charles. "L'expression du verbe de la prière dans le 'carmen' Latin archaïque." In *Recherches sur les Religions de l'Antiquité Classique*. Hautes Études du Monde Gréco-Romain 10, 395-403. Geneva 1980.

(1981a). "Aspects Epiques de la premiere decade de Tite-Live: le rituel de la *devotio*." In *Colloque L'Épopée Greco-Latine et ses prolongements européens*. Calliope II. ed. R. Chevallier. 33-44 Paris 1981.

(1981b). "L'expression du délit dans le rituel archaïque de la prière." In *Le délit religieux dans la cité antique*. Collection de l'École Française de Rome 48, 9-20. Rome 1981.

Halkin, Léon Ernest. *La supplication d'action de graces chez des Romains*. Bibliothèque de la Faculté de Philosophie et Lettres de l'Université de Liège. Fascicle 128. Paris 1953.

Hall, Alan. "New Light on the Capture of Isaura Vetus by P. Servilius Vatia." In *Akten des VI. internationalen Kongresses für griechische und lateinische Epigraphik* 17 (1973) 568-571. Munich 1972.

Hanson, J. A. "Plautus as a Source Book for Roman Religion." *TAPA* 90 (1959) 48-101.

Harris, William V. *War and Imperialism in Republican Rome 327-70 B.C.* Oxford 1979.

Häussler, Reinhard. *Das historische Epos der Griechen und Römer bis Vergil: Studien zum historischen Epos der Antike*. Part 1. *Homer zu Vergil*. Bibliothek der klassischen Altertumswissenschaften. Series 2. Vol. 59. Heidelberg 1976.

Henzen, Wilhelm. *Acta Fratrum Arvalium quae supersunt*. Berlin 1874; reprint 1967.

Herrmann, Peter. *Der römische Kaisereid*. *Hypomnemata* 20. Göttingen 1968.

Heurgon, J. *Trois études sur le 'ver sacrum.'* Brussels 1957.

Highet, Gilbert. *The Speeches in Vergil's Aeneid.* Princeton 1972.

Hirzel, Rudolf. *Der Eid.* Leipzig 1902; reprint, New York 1979.

Janssen, L. F. "Some Unexplored Aspects of *Devotio Deciana.*" *Mnemosyne*, Series 4, 34 (1981) 357-381.

Jeanneret, René. *Recherche sur l'hymme et la prière chez Virgile: Essai d'application de la méthode d'analyse tagmémique à des textes littéraires de l'antiquité.* Collection d'Études Linguistiques. No. 11. Brussels 1973.

Jocelyn, H. D. "Ancient Scholarship and Virgil's Use of Republican Latin Poetry. I." *CQ* 14 (1964) 280-295.

"Varro's *Antiquitates Rerum Divinarum* and Religious Affairs of the Late Republic." In *Bulletin of the John Rylands U. Library*, No. 65, 148-205. Manchester 1982.

Johnson, W. R. *Darkness Visible: A Study of Vergil's Aeneid.* Berkeley 1976.

Kajanto, Iiro. *God and Fate in Livy.* Annales Universitatis Turkuensis, Series B. Vol. 64. Turku 1957.

Kaser, Max. *Das römisches Privatrecht.* Vol. 1. *Handbuch der Altertumswissenschaft. Abt.* 10. Munich^2 1971.

Kaster, Robert. *Guardians of Language.* Berkeley 1988.

Kleinknecht, Hermann. *Die Gebetsparodie in der Antike.* Stuttgart 1937; reprint 1967.

Knauer, Georg Nikolaus. *Die Aeneis und Homer. Studien zur poetischen Technik Vergils mit Listen der Homerzitate in der Aeneis. Hypomnemata* 7. Göttingen 1964.

Köhm, Joseph. *Altlateinische Forschungen.* Leipzig 1905.

Köves-Zulauf, Thomas. *Reden und Schweigen: Römische Religion bei Plinius Maior.* Munich 1972.

Kretzer, Maximillian. *De Romanorum Vocabulis Pontificalibus.* Diss. Halis Saxonum 1903.

Landgraf, Gustav. ed. *Historische Grammatik der lateinischen Sprache.* Leipzig 1903.

Latte, Kurt. *Römische Religionsgeschicte. Handbuch der Altertumswissenschaft. Abt.* 5. *T.* 4. Munich 1960.

Le Gall, J. "*Evocatio.*" 519-524. *Mélanges J. Heurgon*, Vol. 1. Paris-Rome 1976.

Lehr, Heinrich. *Religion und Kult in Vergils Aeneis.* Diss. Giessen 1934.

Liebeschütz, J. H. W. G. *Continuity and Change in Roman Religion.* Oxford 1979.

Linderski, Jerzy. "Notes on CIL I^2 364." In *Parola del Passato* 58 (1958) 47-50.

"Rome, Aphrodisias and the *Res Gestae*: The *Genera Militiae* and the Status of Octavian." *JRS* 74 (1984) 74-80.

(1986a). "The Augural Law." *Aufstieg und Niedergang der Römischen Welt*, 2146-2312. Berlin 1986.

(1986b). "Watching the Birds: Cicero the Augur and the Augural *Templa.*" *CP* 81 (1986) 330-340.

Link, Guilelmus. *De Vocis "Sanctus" Usu Pagano: Quaestiones Selectae.* Diss. Regimonti 1910.

Löfstedt, Einar. *Syntactica: Studien und Beiträge zur historischen Syntax des Lateins.* 2 vols. Lund 1928-33; reprint 1942.

Luce, T. J. *Livy: The Composition of His History.* Princeton 1977.

Lundström, Vilelmus. "Nya Enniusfragment." *Eranos* 15 (1915) 1-24.

Lyne, R. O. A. M. *Ciris: A Poem Attributed to Vergil.* Cambridge 1978.

McCarthy, Dennis J. *Treaty and Covenant: A Study in Form in the Ancient Oriental Documents and in the Old Testament.* Rome^2 1978.

McDonald, A. H. "The Style of Livy." *JRS* 47 (1957) 155-172.

Marquardt, Joachim. *Römische Staatsverwaltung. Handbuch der römischen Alterthümer.* Vol. 4-6. Leipzig^2 1873-1877.

Martinez, David G. *A Greek Love Charm from Egypt. (P. Mich. 757).* Atlanta, Georgia 1991.

Michels, Agnes K. *The Calendar of the Roman Republic.* Princeton 1967.

Mitford, T. B. "A Cypriot Oath of Allegiance to Tiberius." *JRS* 50 (1960) 74-79.

Momigliano, A. Review of A. N. Sherwin-White, *The Roman Citizenship. JRS* 31 (1941) 158-165.

Morani, Moreno. "Dal lessico religioso latino." *Aevum* 57 (1983) 44-50.

Müller, C. F. W. "Zu Plautus." *RhM* 54 (1899) 528-529.

Nicholson, F. W. "The Use of *Hercle* (Mehercle), *Edepol* (Pol), *Ecastor* (Mecastor) by Plautus and Terence." *HSCP* (1893) 99-103.

Nock, A. D. "A Feature of Roman Religion." In *Essays on Religion and the Ancient World*, 481-492. Cambridge, Massachusetts 1972. Previously published in *HTR* 32 (1939) 83-96.

Norden, Eduard. *Agnostos Theos: Untersuchungen zur religioser Rede.* Stuttgart 1913; reprint 1971.

Ennius und Vergilius. Leipzig 1915.

"Aus altrömischen Priesterbüchern." *Acta Reg. Societas Humaniorum Litterarum Lundensis*. Vol. 29. Lund 1939.

North, John A. "Conservatism and Change in Roman Religion." *PBSR* (1976) 1-12.

Ogilvie, R. M. *A Commentary on Livy: Books 1-5*. Oxford 1965.

The Romans and their Gods in the Age of Augustus. New York 1969.

Packard, David W. *A Concordance to Livy*. Cambridge, Massachusetts 1968.

Palmer, L. R. *The Latin Language*. London 1954.

Palmer, Robert E. A. *The Archaic Community of the Romans*. Cambridge, Massachusetts 1970.

Roman Religion and Roman Empire: Five Essays. Philadelphia 1974.

Paladino, Ida. *Fratres Arvales*. Rome 1989.

Parpola, Simo and Watanabe, Kazuko. edd. *Neo-Assyrian Treaties and Loyalty Oaths. State Archives of Assyria*. Vol. 2. Helsinki 1988.

Pasoli, Aelius. ed. with comm. *Acta Fratrum Arvalium.* Studi e Ricerche 7. Bologna 1950.

Pease, Arthur Stanley. ed. with comm. *M. Tulli Ciceronis de Divinatione.* University of Illinois Studies in Language and Literature. Urbana, Book 1, Vol. 6, No. 2, 3, 1920; Book 2, Vol.8, No. 2, 3, 1923.

ed. with comm. *M. Tulli Ciceronis de Natura Deorum.* Vol. 1. Cambridge, Massachussetts 1955; reprint Darmstadt 1968.

Peruzzi, Emilio. "La Lamina dei Cuochi Falischi." *Atti e Memorie dell'Accademia Toscana di Scienze e Lettere la Colombaria* 31 (1966) 115-162.

Peter, Hermann. *Historicorum Romanorum reliquiae.* Vol. 1. Leipzig 1906.

Phillips, Jane. *Official Notices in Livy's Fourth Decade: Style and Treatment.* Unpublished dissertation, University of North Carolina at Chapel Hill 1969.

Pighi, Ioannes Baptista. *De Ludis Saecularibus Populi Romani Quiritium.* Amsterdam[2] 1965.

Pinsent J. "Livy 6.3.1 (*caput rei Romanae*), Some Ennian Echoes in Livy." *LCM* 2 (1977) 13-18.

Pisani, Vittore. *Manuale Storico della Lingua Latina.* Vol. 3. *Testi Latini Arcaici e Volgari.* Turin 1960.

Prugni, G. "*Quirites.*" *Athenaeum,* Ser. 2. 65 (1987) 127-161.

Rawson, Elizabeth. "Scipio, Laelius, Furius, and the Ancestral Religion." *JRS* 63 (1973) 161-174.

Rich, J. W. *Declaring War in the Roman Republic in the Period of Transmarine Expansion.* Collection Latomus. Vol. 149. Brussels 1976.

Ryberg, Inez Scott. *Rites of the State Religion in Roman Art.* MAAR 22. Rome 1955.

Schilling, Robert. *La religion romaine de Vénus depuis les origines jusqu'au temps d'Auguste.* Bibliothèque des Écoles françaises d'Athènes et de Rome. Fascicle 178 (1954). Paris[2] 1982.

"The Roman Religion." *Historia Religionum* I. C. J. Bleeker and G. Widengren, edd. Leiden (1969) 442-494.

Seidl, Erwin. *Der Eid im ptolemäischen Recht*. Munich 1929.

Secknus, Georg. *Untersuchungen zu religiösen Formeln und sonstigen Stellen religiösen Inhalts in den Komödien des Terenz*. Diss., Erlangen 1927.

Shatzman, Israel. "Religious Rites in Virgil's Writings." *SCI* I (1974) 47-63.

Sherwin-White, A. N. *The Letters of Pliny*. Oxford 1966.

Siegert, Hans. "Lat. *esse* und *adesse* als Bewegungsverba." *MH* 9 (1952) 182-191, 256.

Sittl, C. *Gebärden der Griechen und Römer*. Leipzig 1890.

Skutsch, Otto. *Studia Enniana*. London 1968.

ed. with comm. *The Annals of Q. Ennius*. Oxford 1985.

Stacey, S.G. "Die Entwickelung des Livianisches Stiles." *Archiv für lateinische Lexikographie und Grammatik* 10 (1898) 17-82.

Steinwenter. *"Ius iurandum." RE* 10.1 (1918) 1253-1254.

Strzelecki, L. *C. Atei Capitonis Fragmenta*. Wratislavia 1960.

"Naevius and the Roman Annalists." *Rivista di filologia e d'instruzione classica* 91 (1963) 440-458.

Stübler, Gerhard. *Die Religiosität des Livius*. Stuttgart 1941.

Sudhaus, S. "Lautes und leises Beten." *ARW* 9 (1906) 185-200.

Swoboda, Michal and Danielewicz, Jerzy. "Modlitwy i inne partie liryczne w 'Eneidzie' Wergiliusza." *Symbolae Philologicae Poznan* 2 (1975) 57-88.

Syme, R. *Roman Revolution*. Oxford 1939.

Sallust. Berkeley 1964.

Szantyr, Anton. "Über einige Fälle der semantischen Attraktion im Lateinischen." *Gymnasium* 78 (1971) 1-47.

Taübler, Eugen. *Imperium Romanum*. Vol. 1. *Die Staastverträge und Vertagsverhältnisse*. 1913; reprint Rome 1962.

Thulin, Carl Olof. *Italische Sacrale Poesie und Prosa*. Berlin 1906.

Toutain. "*Votum.*" *Daremberg-Saglio.* Vol. 5. 969-978.

Tränkle, H. "Beobachtungen und Erwägungen zum Wandel der livianische Sprache." WS 81 (1968) 103-152.

Ullman, B.L. "History and Tragedy." *TAPA* 73 (1942) 25-53.

Ullmann, Ragnar. "La prose métrique de l'ancienne historiographie romaine." *SO*. Part I (1932) 72-76. Part II (1933) 57-69.

Vahlen, Johannes. ed. *Ennianae Poesis Reliquae.* Leipzig 1903; reprint Amsterdam 1963.

Verbrugghe, Gerald P. "On the Meaning of *Annales*, on the Meaning of Annalist." *Philologus* 133 (1989) 192-230.

Versnel, H. S. "Two Types of Roman Devotio." *Mnemosyne*, Series 4, 29 (1976) 365-410.

"Religious Mentality in Ancient Prayer." In *Faith, Hope, and Worship: Aspects of Religious Mentality in the Ancient World*, ed. by Versnel, 1-64. Leiden 1981.

Wachsmuth, D. "Weihungen." 5.1355-1359. *Kl. Pauly.* Stuttgart 1975.

Wagenvoort, Hendrik. *Roman Dynamism: Studies in Ancient Roman Thought, Language and Custom.* Oxford 1947.

"*De Deae Veneris Origine.*" *Mnemosyne*, Series 4, 17 (1964) 47-77.

"*Orare, precari.*" In *Pietas: Selected Studies in Roman Religion*, 197-209. Leiden 1980. Previously published in *Verbum. Essays dedicated to H. W. Obbink*, 101-111. Utrecht 1964.

Walbank, F. W. *A Historical Commentary on Polybius.* Oxford 1957.

Walsh, P. G. *Livy: his Historical Aims and Methods.* Cambridge 1961.

Watson, Alan. *The Law of Obligations in the Later Roman Republic.* Oxford 1965.

Watson, Lindsay. *Arae: The Curse Poetry of Antiquity.* Leeds 1991.

Weinstock, Stefan. *Divus Iulus.* Oxford 1971.

Wissowa, Georg. *Religion und Kultus der Römer. Handbuch der klassischen Altertumswissenschaft.* 5.4. Munich[2] 1912.

Zieske, Lothar. *Felicitas: Eine Wortuntersuchung. Hamburger Philologische Studien.* No. 23. Hamburg 1972.

Zwierlein, Otto. "Weihe und Entrückung der Locke der Berenike." In *RhM* 130 (1987) 274-290.

Indices

I. Index Rerum

2. Index Verborum
Latin

Greek

3. Index Locorum
Literary Sources

Inscriptions and Papyri